UTLEON3: Exploring Fine-Grain Multi-Threading in FPGAs

T0181762

Martin Daněk • Leoš Kafka • Lukáš Kohout
Jaroslav Sýkora • Roman Bartosiński

UTLEON3: Exploring Fine-Grain Multi-Threading in FPGAs

 Springer

Martin Daněk
Signal Processing
ÚTIA AV ČR, v.v.i.
Pod Vodárenskou věží 1143/4
Praha 8
Czech Republic

Lukáš Kohout
Signal Processing
ÚTIA AV ČR, v.v.i.
Pod Vodárenskou věží 1143/4
Praha 8
Czech Republic

Roman Bartosiński
Signal Processing
ÚTIA AV ČR, v.v.i.
Pod Vodárenskou věží 1143/4
Praha 8
Czech Republic

Leoš Kafka
Signal Processing
ÚTIA AV ČR, v.v.i.
Pod Vodárenskou věží 1143/4
Praha 8
Czech Republic

Jaroslav Sýkora
Signal Processing
ÚTIA AV ČR, v.v.i.
Pod Vodárenskou věží 1143/4
Praha 8
Czech Republic

ISBN 978-1-4899-9570-4   ISBN 978-1-4614-2410-9 (eBook)
DOI 10.1007/978-1-4614-2410-9
Springer New York Heidelberg Dordrecht London

Foreword

Processors and memories are two unavoidable sub-parts of any computing system, as they are at the heart of the ability of the system to compute. Until the turn of the twenty-first century, system engineers using these components as building blocks could assume ever increasing performance gains, by just substituting any of these components by the next generation of the same. Then they ran into two obstacles. One was the memory wall, i.e. the increasing divergence between the access time to memory and the execution time of single instructions. The second was the sequential performance wall, i.e. the increasing divergence in single processors between performance gains by architectural optimizations and the power-area cost of these optimizations. A third potential wall is now profiling on the horizon: the end of Moore's law, or more precisely a potential limit on the maximum number of silicon CMOS–based transistors per unit of area on chip. These obstacles are the stretch marks of a speculation bubble ready to burst. Unless solutions are devised within a decade, the economy of the IT industry of applications, currently based on an expectation of future cheap performance increases in computing systems, may need a serious, globally disruptive reform.

The responsibility to rescue the computing industry falls mostly in the hands of computer engineers, foremost computer architects and systems engineers. An essential step for architects is to find new ways to arrange CMOS logic into different, more efficient combinations of processors and memories. This book is one of the results of a broader initiative taken by the Apple-CORE project, funded by the European Commission, where the expertise of seven international partners was brought together to co-design a new generation of processing tools and components based on hardware multithreading and hardware-supported concurrency management over many cores on chip. A cornerstone of Apple-CORE's scientific and technical output, the UTLEON3 processor demonstrates on FPGAs how novel levels of processing efficiency can be achieved on simple RISC cores with microthreading, paving the way for larger and scalable many-core designs.

However, devising new components may imply invalidating previous assumptions about component semantics. This is the price to pay to take a step forward and break free of the current limitations. The Apple-CORE perspective, for example, limits the use of interrupt-driven preemption, and thus takes away from the hands of software implementers the responsibility of organizing the activation and scheduling of threads. This is why standalone innovation is not sufficient; the *outer question* of innovation, namely *how to ensure that the innovation will be understood, reused and applied*, must also be addressed. And this is where this book shines: by provisioning modular VHDL source code for UTLEON3 under an open license, together with a thorough discussion of the design's internals, motivations and possible applications, the authors have smoothed the learning curve towards understanding and using their design, and demonstrated the tractability of their proposal.

The readers who will most benefit from this monograph are likely newcomers to computer architecture and systems engineering. By relying on FPGA technology and open, modular components, this in-depth review of UTLEON3 both delivers an approachable insight in the future of architecture research and provides an educative and inspiring foundation to build new generations of single-core and multi-core chips.

Amsterdam, The Netherlands   Chris Jesshope
                                                                    Raphael 'kena' Poss

Preface

The UTLEON3 processor has been derived from the LEON3 processor distributed in the GRLIB package by Aeroflex-Gaisler. The main motivation for this work was the evaluation of the performance and hardware cost of the microthreaded computing model in a single core as the first step to designing novel multi-core systems with better scaling properties. The time we spent researching this topic with our colleagues from the Apple-CORE project belongs to the most inspiring and productive periods in our lives as the project created a rare opportunity for interaction between the 'software' and 'hardware' people.

This book presents the design, implementation and evaluation of instruction set extensions for fine-grain multithreading implemented in UTLEON3. The first part of the book defines new processor instructions for thread management in a way compatible with the existing SPARC V8 opcodes, proposes the necessary hardware extensions to LEON3 and describes them on a functional level.

The second part describes implementation details of the implemented architectural extensions that are required to execute microthreads both in the processor pipeline and in specialized hardware accelerators. We have tried to describe the structure of the new blocks in a way that would provide guidance to the actual VHDL sources of the blocks without going into unnecessary details.

The description is accompanied with an analysis of performance gains; these were evaluated by comparing the cycle count of microthreaded assembler programs, executed on UTLEON3 in the microthreaded mode, to reference legacy assembler programs, coded using the standard SPARC V8 instructions and executed on LEON3 and UTLEON3 in the legacy mode.

The main text is supplemented with appendices that provide additional information on LEON3 together with a scheduling example and resource requirements for UTLEON3. As the book is released in parallel with the UTLEON3 sources, a tutorial is also provided that gives instructions on the basic set-up of the VHDL package with the UTLEON3 sources.

We hope you will enjoy the book and UTLEON3.

Praha, Czech Republic   Martin Daněk
                                                                                      Leoš Kafka
                                                                                    Lukáš Kohout
                                                                                  Jaroslav Sýkora
                                                                               Roman Bartosiński

Acknowledgements

This book is one of the results of a nearly 4-year research carried out in the project Apple-CORE. Apple-CORE was funded in the seventh Framework Programme of the European Commission (Project No. FP7-ICT-215216), and also by the Czech Ministry of Education (Project No. 7E08013). For more information about the Apple-CORE project see [17].

In the broader context, the book builds on a 10-year tradition of embedded systems research at the Department of Signal Processing, Institute of Information Theory and Automation of the Academy of Sciences of the Czech Republic (ÚTIA AV ČR, v.v.i.). We gladly acknowledge the support from the Institute that enabled us to delve into this work.

The information in this book is presented to the best of our knowledge, still you are encouraged to subject it to your own critical thinking and common sense before applying it.

Contents

Acronyms

ADSL	assymetric digital subscriber line
AHB	advanced high-performance bus
AMBA	advanced microcontroller bus architecture
BCE	basic computing element
BlockRAM	RAM implemented in a special ASIC block in an FPGA
BRAM	BlockRAM
Cacheline	sixteen 32-bit words that are fetched in one fetch transaction from an aligned address in UTLEON3
CC	clock cycle
CL	cacheline
DCT	discrete cosine transform
DE	decode stage
DMA	direct memory access
DP RAM	dual-port RAM
DSP	digital signal processing
D-Cache	data cache, dcache
EX	execute stage
FE	fetch stage
FGS	family global storage
FID	family identifier
FIR	finite impulse response
FPGA	field-programmable gate array
FTT_A_WIDTH	width of the address field in the family table
GRLIB	Gaisler Research VHDL library with IP cores and sample designs that use them
holdn	signal that halts the integer pipeline
HW	hardware
HWFAM	hardware accelerator for families of threads
I-Cache	instruction cache, icache
I/O	input/output
IU3	integer unit (pipeline) in LEON3

JCU	job control unit
L3	legacy mode
RA	register access stage
RAM	random access memory
RAU	register allocation unit
regfile	register file
RUC	register update controller
MA	memory access stage
MUL	integer multiplier, integer multiplication
SPARC	scalable processor architecture
SW	software
td	thread dependent register (in)
tg	thread global register (in)
TID	thread identifier
tl	thread local register (scratchpad)
TLS	thread local storage
TMT	thread mapping table
ts	thread shared register (out)
TT	trap type
TT_A_WIDTH	width of the address field in the thread table
UT	microthreaded mode
UTGRLIB	microthreaded GRLIB
UTLEON3	microthreaded LEON3
WB	writeback stage
XC	exception stage

Part I
Programming Interface

Chapter 1
Introduction

The current silicon technology has reached its limits in maximum operating frequency, thus it has become more important to improve the organization of computation to gain further performance improvements. One area for improvements is to eliminate computation stalls on execution of long-latency operations. In processors these situations can be eliminated by organizing the computation in several independent threads, and switching the context to threads that are ready for execution in situations when processor execution would be stalled otherwise.

As the silicon area becomes cheaper as a consequence of the Moore's law, it has become viable to extend processors to support in hardware execution of multiple threads in one processor or in a multiprocessor cluster. Two significant examples are the SUN Microsystems OpenSPARC T1/T2 and the MIPS MT processors. OpenSPARC T1/T2 is an open-source version of the UltraSPARC T1/T2 [13, 16]; T1 has been ported to the Xilinx FPGAs, while MIPS MT [12] is a commercial processor available as an ASIC. We believe the architecture complexity of the open-source OpenSPARC T1/T2 is too high for embedded applications, which is due to their primary domain in server and desktop computing. Also the context switch time for T1/T2 is high, about 1,000 clock cycles. We do not know of any other multithreaded processor available in the source code to the design community.

We have designed and implemented instruction set extensions for the simpler LEON3 SPARCv8 processor [4] suitable for embedded applications. The reasons why we have chosen LEON3 are its availability in VHDL under GPL, the availability of the standard GNU toolchain for LEON3 together with linux ports, and last but not least the development support we received from Gaisler Research (now Aeroflex-Gaisler) within the project Apple-CORE funded by the European Commission.

This book describes the resulting architecture of the modified LEON3 processor that we call UTLEON3, and the impact of the architectural improvements on the processor performance and hardware requirements. The goal has been twofold: first, we wanted to implement in silicon machine-level instructions that were identified as necessary for efficient microthreading – a variant of multithreading with fast context switching [11] to have a true picture how these extensions are expensive in terms of

M. Daněk et al., *UTLEON3: Exploring Fine-Grain Multi-Threading in FPGAs*,
DOI 10.1007/978-1-4614-2410-9_1, © Springer Science+Business Media, LLC 2013

silicon (compared to optimistic performance results from the functional simulation). Second, as the first step to full multi-core implementation of microthreading, we wanted to show that we could achieve an efficient single-core implementation of these extensions by outperforming the original LEON3 processor by better handling of memory and execution latencies.

Still the biggest promise of microthreading remains to be proven in silicon – whether, given the current technological limitations, a multi-core system based on microthreading can achieve linear scalability with respect to the number of processing cores as indicated in functional simulations.

1.1 Microthreading

Microthreading is a multithreading variant that decreases the complexity of context management. The goal of microthreading is to tolerate long-latency operations (LD/ST and multi-cycle operations such as floating-point) and to synchronize computation on register access. An overview of multithreading is provided in [18].

In a simple case the context can be represented by the program counter and by window pointers to the register file. Microthreading has been developed both on the assembler and C levels. The basic conceptual unit is a family of threads that share data and implement one piece of computation. In a simple view one family corresponds to one for-loop in the classical C; in microthreading each iteration (each thread) of a hypothetical for-loop (represented by a family of threads) is executed independently according to data dependencies. A family is synchronized on termination of all its threads. More details on microthreading can be found in [9–11].

A possible speedup generated by microthreading comes from the assumption that while one thread is waiting for its input data, another thread has its input data ready and can be scheduled ideally in zero clock cycles and executed. Another assumption is that load and store operations themselves need not be blocking since the real problem arises just when an operation accesses a register that does not contain a valid data value. Finally, the thread management logic is considered simple enough to fit in the processor hardware reasonably well in the current technologies.

The hardware requirements of microthreading are: use of a self-synchronizing register file (i-structures, [1]), register states to be managed autonomously in the register file, pipeline stalls prevented by context switch in hardware, and thread status and context switch managed autonomously in a hardware thread scheduler.

The microthreading support on the machine level is represented by the following assembler instructions:

- *launch* – switches the processor from the legacy mode (user or protected) to the microthreaded mode.
- *allocate* – allocates a family table entry, needed to create a family of threads.

- *set...* – several instructions that fill in the allocated family table entry with parameters required by the *create* instruction.
- *create* – creates (a family of) threads based on a family table entry.
- *.registers* – a pseudoinstruction at the beginning of the thread code that specifies the number of global, local and shared registers needed by a thread (integer and floating-point specified separately).

Furthermore, each 32-bit instruction word is extended by another two bits that together encode an instruction modifier for thread scheduling. The defined modifiers are:

- *cont* – continue thread execution (default),
- *swch* – switch the context to another thread, e.g. on memory load to prevent possible pipeline stall,
- *end* – end thread execution, i.e. the thread ends at this instruction.

The format of assembler instructions has been extended by a field delimited by a semicolon that may contain an explicit instruction for the scheduler. If the field is missing, *cont* is assumed by default.

```
clr %r2
ld [%r1 + %g0], %r3 ; swch
add %r3, %g0, %r4 ; end
```

To keep the 32-bit organization of the memory system in SPARCv8, 2-bit extensions or modifiers for groups of 15 instructions are grouped in one 32-bit instruction word that is located at the beginning of each instruction cache line (one cache line is formed by 16 words). The first word of each cacheline is skipped in the microthreaded mode (explained in more detail later in the text). The instruction cache organization is shown in Fig. 1.1.

Microthreading relies on the use of a self-synchronizing register file based on the *i-structures* [1]. To implement the i-structures registers are extended with states. A register can be:

- *empty* – on power-on reset,
- *pending* – a memory load operation has been requested and no thread has accessed the register since,
- *waiting* – a memory load operation has been requested and a thread has accessed the register since,
- *full* – the register contains valid data.

In the micro-threading model a pending register can be accessed by at most one thread – either by the thread that initiated the pending data update, or by its direct sibling (only unidirectional data dependencies between direct sibling threads are allowed in microthreading to prevent deadlock).

A sample program execution is shown in Fig. 1.2. The processor starts in the legacy mode on power-on reset, then it switches to the microthreaded mode.

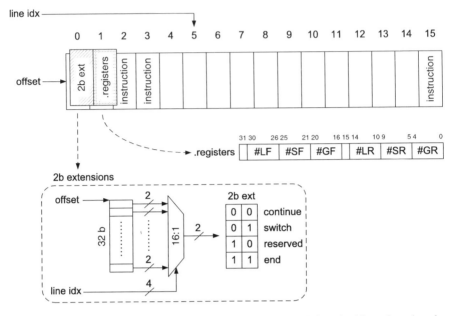

Fig. 1.1 Organization of the instruction cache in the microthreaded mode. 16 words = 1 cache line

The parent thread gets synchronized with the children threads by reading the register %*l2* (this can be omitted when thread synchronization is not required). On completion of all microthreads the processor switches back to the legacy mode.

1.2 UTLEON3

UTLEON3 is a microthreaded processor derived from LEON3. LEON3 is a successor of the ERC and LEON2 processors, all developed by Jiri Gaisler and later Gaisler Research for the European Space Agency. LEON3 is a RISC processor based on the SPARC V8 specification, standardized by IEEE as Std 1754–1994. It is distributed under the GNU GPL scheme in a freely available *GRLIB* package with all necessary VHDL source codes. The level of available documentation is sufficient for building complex systems on chip (similar to commercial products such as Xilinx EDK or Altera SOPC Builder) featuring one or more LEON3 cores with the necessary peripherals. The UTLEON3 processor is distributed under the GNU GPL scheme in a *UTGRLIB* package derived from the *GRLIB* package.

The book is organized in three parts. The first part focuses on the programming interface to microthreading in UTLEON3. After a presentation of background information on the *GRLIB* library and LEON3, the new microthread management instructions are described, followed by a functional description of the LEON3

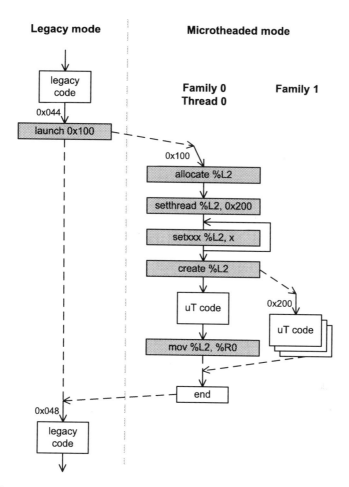

Fig. 1.2 Program flow

processor. The first part is concluded with a few programming examples that demonstrate the approach to writing microthreaded assembler programs.

The second part describes the implementation details. It starts with describing the structure of the key new blocks, including hardware acceleration of families of microthreads and interrupt handling, and evaluates efficiency of the microthreaded model in terms of resource requirements, operating frequency and execution efficiency.

Supplementary information, including an analysis of resource requirements of UTLEON3 for an FPGA and ASIC implementations can be found in the appendices.

Chapter 2
The LEON3 Processor

This section contains a brief description of the LEON3 SPARC V8 processor implementation developed by Gaisler Research, with an emphasis on information relevant to the derived UTLEON3 microthreaded processor. The LEON3 processor is part of the *GRLIB* package that is under continuous development; the following description is based on *GRLIB* Version 1.0.17. The description provided in this section is organized in a top-down manner, starting with a view of the whole IP library *GRLIB* and going down to the LEON3 integer pipeline.

2.1 The GRLIB Library

GRLIB is a library of VHDL source codes of IP cores for designing a complete system on chip centered around the LEON3 processor [4,5]. The library is structured into directories that group IP cores according to the contributor's company. There are more subdirectories with complete IP cores organized in packages in the directories. These directories contain VHDL source codes of packages for simulation and synthesis and files with package configurations. A package configuration is propagated to VHDL entities via generics. The library is managed by an automated tool based on *GNU make* that manages configurations and compilation of packages for simulation and synthesis.

2.1.1 Two-Process Coding Style

The majority of the VHDL code in the GRLIB library is coded using the two-process method developed by Gaisler [2], where one process forms the combinational part of the logic and uses variables to compute the results, and the other process implements register updates. All buses are coded as records; this reduces the coding effort when adding or removing interface signals.

M. Daněk et al., *UTLEON3: Exploring Fine-Grain Multi-Threading in FPGAs*, DOI 10.1007/978-1-4614-2410-9_2, © Springer Science+Business Media, LLC 2013

The advantage of this coding method is shorter simulation time, but the price paid is lower readability of the code for designers not familiar with this coding style (namely due to C-like sequential dependencies between variable assignments in the combinational process in contrast to the usual parallel nature of VHDL signal assignments). Newer versions of *doxygen* can be used to extract documentation from the VHDL sources.

2.1.2 GRLIB Directory Structure

The LEON3 processor consists of several parts. The main parts of the processor implementation are placed in the *$GRLIB/lib/gaisler/leon*3 directory. VHDL files with the SPARC V8 instruction codes and support functions for instruction disassembly during simulation are placed in the *$GRLIB/lib/grlib/sparc* directory. User designs with top-level files that instantiate complete SoC designs are located in the *$GRLIB/designs* directory.

2.2 LEON3

LEON3 is an implementation of the SPARC V8 architecture with several specific features. Figure 2.1 shows a block diagram of the LEON3 processor with optional parts (e.g. DIV32, MUL32, DSU, MMU).

The minimal configuration of the LEON3 processor consists of an integer pipeline, 3-port register file, instruction cache and data cache. A more powerful configuration can contain a memory management unit, 32-bit integer multiplication and division unit, 16-bit integer multiply-accumulate unit, debug interface and trace buffers according to its configuration. LEON3 has two extension interfaces, usually used by a floating-point unit and a co-processor.

More information about the SPARC V8 architecture and the LEON3 implementation are in [3, 5, 7, 15]. One of our initial design requirements was that the new UTLEON3 processor be binary compatible with the original LEON3 processor. To better understand the design decisions we took for UTLEON3 we describe the relevant parts of LEON3 in the following text.

2.2.1 Pipeline Operation

The LEON3 processor execution is controlled by the integer pipeline. The pipeline is implemented in seven independent stages that use delayed branches and several forwarding paths to achieve high performance. A schematic view of the pipeline is shown in Fig. 2.2.

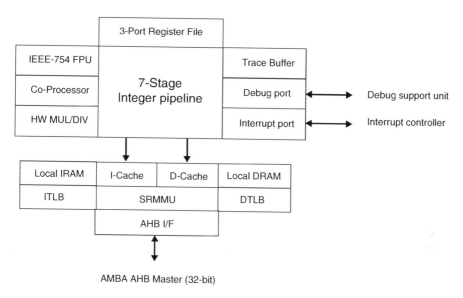

Fig. 2.1 Structure of the LEON3 processor

The integer pipeline is coded in the file *iu3.vhd*. The state of the whole *iu3* pipeline is captured in one signal, named *r*. This signal is a record of records, where each level-1 record corresponds to each pipeline stage. The level-2 records are tailored to the requirements of each pipeline stage, some items propagate through several pipeline stages. The structure of the pipeline data structure is included in Appendix A.

2.2.2 Instruction Set

Instructions in the LEON3 processor are encoded in three 32-bit formats. Instruction formats and an excerpt from the instruction table is included in Appendix B. The LEON3 integer pipeline implements the full SPARC V8 standard, including hardware multiply and divide instructions, and in addition it implements hardware multiply-accumulate instructions.

Instructions can be divided in six categories: load/store, integer arithmetic, control transfer, read/write control registers, floating-point operations and co-processor operations. There are several free places in the *Format 3* (op = 2) instruction set table. The load/store instructions can use alternate address spaces defined by the 8-bit field *ASI* in the instruction code. LEON3 uses only a few address spaces, and there are a lot of spaces for possible extensions. There are three empty places for new branch (flow control) instructions.

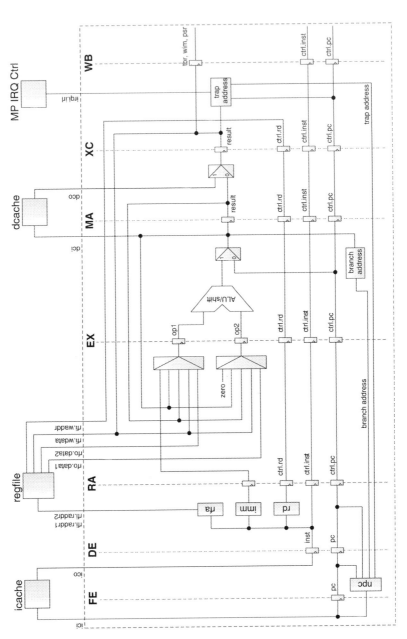

Fig. 2.2 The LEON3 integer pipeline – a simplified view

All instructions can behave differently in the user and privileged modes, and they can cause an exception if they are called in an improper mode. Load/store data from/to an improper address space can cause the invalid memory access exception.

2.2.3 Registers

The LEON3 processor includes all the necessary registers as defined in the SPARC V8 specification. It contains general-purpose registers (the %r registers) and control/status registers (%psr, PC, etc.) as shown in Appendix C.

The general-purpose registers are included in a register file that is connected to the processor instruction pipeline through three ports. The register file is organized as 8 global registers and an implementation dependent number of 16-register sets. Instructions can access all 8 global registers (global), 16 registers of the current window (local, in) and the 8 in registers of the next window (out). The current window is changed only by the SAVE and RESTORE instructions or when entering or returning from a trap. The SPARC V8 specification recommends to implement 8 register sets for compatibility reasons. Four r registers have a special function. The register %r15 (%o7) is used for saving the program counter of a CALL instruction. When a trap occurs, the pc and npc are copied into registers %r17 and %r18 (%l1 and %l2) of the trap's new register window. Register %r0 (%g0) behaves in a specific way; if %r0 is a source operand, the constant value 0 is read. When %r0 is used as a destination operand, the result of the operation is discarded.

2.2.4 Exceptions

LEON3 implements the general SPARC trap model. The table of the implemented traps and their individual priorities is in the GRLIB IP User's Manual [5]. Trapping is enabled by setting the bit ET in processor status register. LEON3 supports single vector trapping (SVT) that can reduce the code size of trap handlers. In the SVT mode all traps use a handler to reset a trap, and the trap type is indicated in the field TT in the register TBR.

2.2.5 Software Tools

The software tools needed for building a microthreaded processor from the LEON3 VHDL source codes are fully described in the GRLIB User's Manual [4]. These tools don't have to be modified.

Programs for the LEON3 processor can be processed using a modified GNU toolchain for the SPARC architecture. The source code of these tools is available on the *Gaisler Research* web site. The minimal setting requires at least the GNU assembler to program the processor.

2.3   From LEON3 to UTLEON3

The internal structure of the UTLEON3 processor is derived from LEON3, but certain parts are different. These are namely:

Instruction decoder – new instructions that support microthreaded execution.

Integer pipeline – new instructions, new program flow control mechanism (thread switch).

Register file – new registers, new register state bits, new register allocation scheme.

Traps – new traps due to error states in the microthreaded mode.

Caches – new cache controllers that support delayed memory accesses (R/W), independent register file updates, and reference counting.

Program flow – a distributed call stack equivalent stored in the register file, thread scheduler and instruction cache.

UTLEON3 maintains the standard SPARCv8 instruction formats and opcode table. New microthread management instructions have been implemented over specific cases of the SPARCv8 RDASR and WRASR instructions.

Chapter 3
Microthreaded Extensions

This chapter defines the microthreaded extensions to LEON3 that can be seen on the assembler level. These are namely the processor identification and program status register, and the new instructions that switch the processor to the microthreaded mode and manage thread creation and execution.

3.1 Definitions

A *thread* is a part of the program that implements one iteration of a hypothetical loop. A *family of threads* is a group of threads that implements computation of one loop. A family is defined by its *index start*, *end*, *step*, and also by its *body* – the program code that executes one thread. A *parent thread* is the wrapper code that created a given family of threads. A *child thread* is one thread that represents one iteration of a hypothetical loop. *Sibling threads* execute simultaneously within one family, i.e. they represent different iterations of a hypothetical loop.

All families that exist in the system are listed in the *family table*. The table contains a pointer to the first thread of each family in the *thread table*. The *thread table* contains all threads that have been created in the system and have not finished yet, that is the threads that are running, ready, waiting and/or suspended in the system. Each thread in a given family is represented by its unique *thread identifier (TID)*. Each family that exists in the system is represented by its unique *family identifier (FID)*.

Thread execution can be *synchronized* both between sibling threads and between the parent and children threads. The synchronization is implemented through register reads in the self-synchronizing register file. In most situations a parent thread will get synced on family completion by reading the register that contains the family return value – the register marked as pending on family creation will get updated on family completion.

In the following simple case a family of threads is equivalent to one unrolled loop, where each thread computes one loop iteration.

M. Daněk et al., *UTLEON3: Exploring Fine-Grain Multi-Threading in FPGAs*, DOI 10.1007/978-1-4614-2410-9_3, © Springer Science+Business Media, LLC 2013

```
$index=$start
do {
  ... thread_body ...
  $index = $index + $step;
} while ($index < $limit);
```

Default values *start=0, limit=0, step=1* implement one pass through the loop, i.e. they will result in the creation of just one thread. An infinite loop can be created by setting *step=0*. The loop index need not reach exactly the *limit* value to terminate the loop. The loop terminates when *index >= limit* (similar to C).

It is also worth mentioning here that the loop control instructions (decrements, branching) are not used in the microthreaded code as these are executed in the hardware thread scheduler. We will return to this later.

3.2   UTLEON3 Program Status Register

Each SPARC CPU type is identified by two fields of the PSR – identification of the implementation (4bits) and implementation version (4bits). UTLEON3 is identified by the implementation version 0xAE (original LEON3 value is 0×3).

```
1  constant VENDOR_APPLECORE   : amba_vendor_type  := 16#AE#;
```

Two new devices are included in the Gaisler Research GRLIB tools: UTLEON3 with device type 0x1, and UTLEON3DSU with device type 0×2.:

```
1  constant APPLECORE_UTLEON3       : amba_device_type  := 16#001#;
   constant APPLECORE_UTLEON3DSU : amba_device_type  := 16#002#;
```

The microthreaded mode is indicated by bit 19 in the PSR (a newly-defined *Microthreaded Mode* bit).

3.3   Switching from the Legacy Mode to the Microthreaded Mode

The processor switches from the legacy to the microthreaded mode on execution of the *launch* instruction (see Fig. 4.2). An address of the first instruction of the microthreaded code is provided as its argument. The *launch* instruction is the last instruction that is processed in the legacy mode; the subsequent legacy instruction is annuled; it will be re-fetched when the processor switches back to the legacy mode. The instruction that will be executed instead is provided by the thread scheduler.

3.4 Switching from the Microthreaded Mode to the Legacy Mode

The processor switches back to the legacy mode when *Thread 0* decodes an instruction with the *END* modifier (see Fig. 4.2).

3.5 New UTLEON3 Instructions

An overview of new instructions together with the relevant part of the opcode is shown in Table 3.1.

NOTE: Instructions marked with an asterisk (*) have not been implemented in UTLEON3 yet.

Table 3.1 Summary of the new UTLEON3 instructions for microthreaded execution

Instruction	R/W	op_uT
launch	W	b0001
allocate	R	b0001
setstart	W	b0010
setlimit	W	b0011
setstep	W	b0100
setblock	W	b0101
setthread	W	b0110
setregg*	W	b1000
setregs*	W	b1001
create	R	b0010
getfid	R	b0100
gettid	R	b0011
break*	W	b1010
kill*	W	b1011

3.5.1 *Launch*

Switch UTLEON3 from the legacy to microthreaded mode.

UTLEON3 syntax:

```
launch %reg_rs1
```

It is defined as a special case of the instruction

```
WRASR  %reg_rs1, %reg_rs2, %ASR20
```

with bit *i* set to '0' and a part of the instruction opcode replaced with a new field *op_uT* that identifies the instruction `launch`; it is equal to `b0001`.

- `%reg_rs1` – address of the microthreaded code starting address.

Instruction format:

op		rd=20			op3				rs1					i	op_uT				unused(zero)							
1	0	1	0	1	0	0	1	1	0	0	0	0	x	x	x	x	x	0	0	0	0	1	0	0	0	0	0	0	0	0

31 30 29   25 24   19 18   14 13 12   9 8   0

Description:

The extended LEON3 processor can execute both the legacy and the microthreaded code. The legacy code is executed without any changes compared to execution on the standard LEON3 processor. The code execution always starts in the legacy mode, and after an initial set up the processor is switched to the microthreaded mode with the `launch` instruction. The instruction passes the starting address address of the microthreaded code to the thread scheduler. When the microthreaded execution finishes, the processor is switched back to the legacy mode. This behaviour is very similar to a subprogram call.

Semantics:

1. Switch the UTLEON3 mode from the legacy mode to the microthreaded mode.
2. Wait until the legacy instructions in the integer pipeline finish.
3. Send the microthreaded code starting address to the scheduler.

Complexity:

O(1)

Affected data structures:

- Family table.
- Thread table.
- Thread scheduler.
- Processor pipeline.

Integration in IU3:

- DE – switch the IU3 into the microthreaded mode, invalidate the subsequent instruction.
- RA – read the microthreaded code start address from the register.
- EX – write the microthreaded code start address to the thread scheduler.

Exceptions:

None.

3.5.2 *Allocate*

Allocate entries in the family table.

UTLEON3 syntax:

```
allocate %reg_rd
```

The instruction is defined as a special case of the instruction

```
RDASR %ASR20, %reg_rd
```

with bit i set to '0', and a part of the instruction opcode replaced with a new field op_uT that identifies the instruction `allocate`; it is equal to b0001.

- `%reg_rd` – index of the allocated family (FID)

Instruction format:

op	rd	op3	rs1=20	i	op_uT	unused(zero)
1 0	x x x x x	1 0 1 0 0 0	1 0 1 0 0	0	0 0 0 1	0 0 0 0 0 0 0 0 0

bit positions: 31 30 29 ... 25 24 ... 19 18 ... 14 13 12 ... 9 8 ... 0

Description:

The instruction returns in a register a pointer to a free entry in the family table and marks this entry as allocated. If no entry is free, the instruction generates an exception.

Semantics:

- If there is a free entry in the family table:
  - Allocate an entry (set the entry status to *ALLOCATED*, also called as *FID valid*).
  - Set the output register to the FID value.
- If there is no free entry in the family table:
  - Generate an exception.

Complexity:

O(1)

Affected data structures:

- Family table
- Destination register

Integration in IU3:

- RA – read empty FID from the family table.
- WB – write the FID to a register.

Exceptions:

None

3.5.3 SetStart

Set the initial index value.

UTLEON3 syntax:

```
setstart %reg_rs1, %reg_rs2
setstart %reg_rs1, imm9
```

The instruction is defined as a special case of the instruction

```
WRASR  %reg_rs1, %reg_rs2, %ASR20
```

with bit i set to:

- '0' – the second operand is a register
- '1' – the second operand is a 9b unsigned immediate value

and a part of the instruction opcode replaced with a new field op_uT that identifies the instruction setstart; it is equal to b0010.

- %reg_rs1 – the FID of the previously allocated family
- %reg_rs2 or imm9 – the value of the family initial index

Instruction format:

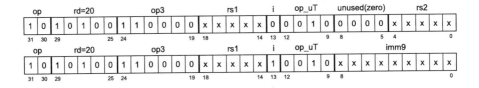

Description:

The instruction sets a parameter required to create a family of threads using the create instruction. It sets the initial index value for a family. A default value is 0.

Semantics:

- Write a value to the *start* field of the specified family table entry.

Complexity:

O(1)

Affected data structures:

• Family table – one table entry.

Integration in IU3:

• RA – read an FID value from a register.
• EX – write a value to the family table.

Exceptions:

The instruction generates an exception if the FID specifies an unallocated family table entry.

3.5.4 SetLimit

Set the final index value.

UTLEON3 syntax:

```
setlimit %reg_rs1, %reg_rs2
setlimit %reg_rs1, imm9
```

is defined as a special case of the instruction

```
wrasr %reg_rs1, %reg_rs2, %reg_rd
```

with bit *i* set to :

- '0' – the second operand is a register
- '1' – the second operand is a 9-bit unsigned immediate value

and a part of the instruction opcode replaced with a new field *op_uT* that identifies the instruction setlimit; it is equal to b0011.

- %reg_rs1 – the FID of the previously allocated family
- %reg_rs2 or imm9 – the value of the family limiting index.

Instruction format:

op	rd=20	op3	rs1	i	op_uT	unused(zero)	rs2
1 0	1 0 1 0 0	1 1 0 0 0 0	x x x x x	0	0 0 1 1	0 0 0 0	x x x x x

31 30 29   25 24   19 18   14 13 12   9 8   5 4   0

op	rd=20	op3	rs1	i	op_uT	imm9
1 0	1 0 1 0 0	1 1 0 0 0 0	x x x x x	1	0 0 1 1	x x x x x x x x x

31 30 29   25 24   19 18   14 13 12   9 8   0

Description:

The instruction sets a parameter required to create a family of threads using the create instruction. It sets the final index value for a family. The default value is 0.

Semantics:

- Write a value to the *limit* field of a specified family table entry.

Complexity:

O(1)

Affected data structures:

- Family table – one table entry.

Integration in IU3:

- RA – read an FID value from a register.
- EX – write a value to the family table.

Exceptions:

The instruction generates an exception if the FID specifies an unallocated family table entry.

3.5.5   SetStep

Set the index increment.

UTLEON3 syntax:

```
setstep %reg_rs1, %reg_rs2
setstep %reg_rs1, imm9
```

is defined as a special case of the instruction

```
WRASR %reg_rs1, %reg_rs2, %ASR20
```

with bit i set to :

- '0' – the second operand is a register
- '1' – the second operand is a 9-bit unsigned immediate value

and a part of the instruction opcode replaced with a new field op_uT that identifies the instruction `setstep`; it is equal to b0100.

- `%reg_rs1` – the FID of the previously allocated family
- `%reg_rs2` or `imm9` – the value of the family index increment

Instruction format:

op		rd=20					op3						rs1					i	op_uT			unused(zero)					rs2				
1	0	1	0	1	0	0	1	1	0	0	0	0	x	x	x	x	x	0	0	1	0	0	0	0	0	0	x	x	x	x	x

31 30 29   25 24   19 18   14 13 12   9 8   5 4   0

op		rd=20					op3						rs1					i	op_uT			imm9								
1	0	1	0	1	0	0	1	1	0	0	0	0	x	x	x	x	x	1	0	1	0	0	x	x	x	x	x	x	x	x	x

31 30 29   25 24   19 18   14 13 12   9 8   0

Description:

The instruction sets a parameter required to create a family of threads using the `create` instruction. It sets the index increment for a family. The default value is 1.

Semantics:

- Write a value to the *limit* field of a specified family table entry.

Complexity:

O(1)

Affected data structures:

- Family table – one table entry.

Integration in IU3:

- RA – read an FID value from a register.
- EX – write a value to the family table.

Exceptions:

The instruction generates an exception if the FID specifies an unallocated family table entry.

3.5.6 *SetBlock*

Set the limit of the number of threads in the thread table that belong to one family of threads.

UTLEON3 syntax:

```
setblock %reg_rs1, %reg_rs2
setblock %reg_rs1, imm9
```

is defined as a special case of the instruction

```
WRASR %reg_rs1, %reg_rs2, %ASR20
```

with bit *i* set to:

- '0' – the second operand is a register
- '1' – the second operand is a 9-bit unsigned immediate value

and a part of the instruction opcode replaced with a new field op_uT that identifies the instruction `setblock`; it is equal to b0101.

- `%reg_rs1` – the FID of the previously allocated family
- `%reg_rs2` or `imm9` – the value of the family block size

Instruction format:

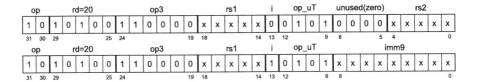

Description:

The instruction sets a parameter required to create a family of threads using the `create` instruction. It sets the maximal number of threads from one family that can be created at one time. The default value is 0; in this case the actual value is calculated as the minimum of the size of the thread table (the number of entries) and the number of threads in the family.
Semantics:

- Write a value to the *block* field of a specified family table entry.

Complexity:

O(1)

Affected data structures:

• Family table – one table entry.

Integration in IU3:

• RA – read an FID value from a register.
• EX – write a value to the family table.

Exceptions:

The instruction generates an exception if the FID specifies an unallocated family table entry.

3.5.7 *SetThread*

UTLEON3 syntax:

```
setthread %reg_rs1, %reg_rs2
setthread %reg_rs1, imm9
```

is defined as a special case of the instruction

```
WRASR %reg_rs1, %reg_rs2, %ASR20
```

with bit *i* set to:

- '0' – the second operand is a register
- '1' – the second operand is a 9-bit unsigned immediate value

and a part of the instruction opcode replaced with a new field *op_uT* that identifies the instruction setthread; it is equal to b0110.

- %reg_rs1 – the FID of the previously allocated family.
- %reg_rs2 or imm9 – the value of the family thread function.

Instruction format:

op		rd=20		op3				rs1		i	op_uT		unused(zero)		rs2	

```
  op       rd=20              op3            rs1       i    op_uT   unused(zero)      rs2
 1 0 | 1 0 1 0 0 1 1 0 0 0 0 | x x x x x | 0 | 0 1 1 0 | 0 0 0 0 | x x x x x
 31 30 29            25 24            19 18      14 13 12    9 8       5 4         0

  op       rd=20              op3            rs1       i    op_uT       imm9
 1 0 | 1 0 1 0 0 1 1 0 0 0 0 | x x x x x | 1 | 0 1 1 0 | x x x x x x x x x
 31 30 29            25 24            19 18      14 13 12    9 8                    0
```

Description:

The instruction sets a pointer to the family thread function.

Semantics:

- If the family with the specified FID is allocated (and not created):
  - Write the value to the thread address field of the specified family table entry.
- If the FID is invalid or the family has already been created:
  - Generate an exception TT_UTERR.

Complexity:

O(1)

Affected data structures:

• Family table – one table entry.

Integration in IU3:

• RA – read an FID value from a register.
• EX – write a value to the family table.

Exceptions:

The instruction generates an exception if the FID specifies an unallocated family table entry.

3.5.8 SetRegs[*]

Set the base addresses of the register windows for children threads (i.e. pointers to the register file).

UTLEON3 syntax:

```
setregg %reg_rs1, %reg_rs2
setregg %reg_rs1, imm9
setregs %reg_rs1, %reg_rs2
setregs %reg_rs1, imm9
```

are defined as a special case of the instruction

```
WRASR %reg_rs1, %reg_rs2, %ASR20
```

with bit *i* set to:

- '0' – the second operand is a register
- '1' – the second operand is a 9-bit unsigned immediate value

and a part of the instruction opcode replaced with a new field op_uT. For the instruction setregg the field is equal to b1000. For the instruction setregs it is equal to b1001.

- %reg_rs1 – the input register with the FID of the previously allocated family.
- %reg_rs2 or imm9 – the input register with a new index to register file for global and shared registers.

Instruction format:

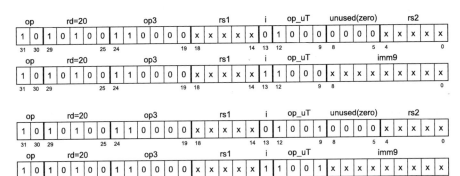

Description:

The instruction sets the base addresses of register windows global and shared registers (i.e. pointers to the register file) for the children threads. The instruction is actually implemented as two separate instructions `setregg` and `setregs` for global and local registers, respectively.

Semantics:

- Write a value to the *regs* and *regs2* field of the specified family table entry.

Complexity:

O(1)

Affected data structures:

- Family table – one table entry.

Integration in IU3:

- RA – read an FID value from a register.
- EX – write a value to the family table.

Exceptions:

The instruction generates an exception if the FID specifies an unallocated family table entry.

Note:

Not implemented.

3.5.9 *Create*

Initiate creation of a family of threads.

UTLEON3 syntax:

```
create %reg_rs2, %reg_rd
```

is defined as a special case of the instruction

```
RDASR   %ASR20, %reg_rd
```

with bit *i* set to '0' and a part of the instruction opcode replaced with a new field *op_uT* that identifies the instruction `create`; it is equal to b0010.

- `%reg_rs2` – the FID.
- `%reg_rd` – the family termination code.

Instruction format:

op		rd						op3					rs1=20		i	op_uT				unused(zero)				rs2				
1	0	x	x	x	x	x	1	0	1	0	0	0	1	0	1	0	0	0	1	0	0	0	0	x	x	x	x	x

31 30 29   25 24   19 18   14 13 12   9 8   5 4   0

Description:

The instruction issues a request for creation of a specified family of threads. The family was allocated with the `allocate` instruction and set with the `set...` instructions. The source register `%reg_rs2` must contain a valid FID. The destination register is set to *PENDING* at the time the instruction is executed. When the family finishes, the destination register is set to *FULL* by the hardware thread scheduler, and it contains the family termination code.

Semantics:

After the `create` instruction is issued, the hardware thread scheduler will:

- Determine the number of family threads from the values *start, limit, step*:
  - $step > 0$ ... increment index, $\#threads = \lceil ((limit - start)/step) \rceil$
- Determine the maximal number of threads executed at once (the size of the block of threads):
  - $block = 0$ ... $block = min(\#threads, \#tids)$
  - $block \neq 0$ ... the size of the block is the value of *block*

- Determine the number of registers assigned to one thread from the first word of the thread code (the assembler .register pseudo-instruction stored at address *thread_starting_address − 0x4*)
- Allocate the required number of registers from the pool of free register blocks.
- Allocate thread IDs in the thread table (at most *#block* threads) sequentially and for each thread perform the following steps:
  - Calculate the thread index.
  - Determine the TID.
  - Initialize the thread registers.
  - Initialize the thread state and link the thread table entries and family table entries for all created threads.

A thread can read its index value in its first local register (*%l0*).

A running thread is suspended when it accesses a *PENDING* register, e.g. a register to be written by other threads, or a destination register of a data load operation (cache miss). This mechanism is also used to synchronize the execution of a parent thread and its children threads.

Complexity:

O(1)

Affected data structures:

- Family table – one table entry, data update.
- Thread scheduler.
- Register file – mark a destination register *PENDING*.

Integration in IU3:

- RA – read an FID value from a register.
- EX – write a value to the family table.
- WB – mark a register with the FID value *PENDING*.

Exceptions:

An exception can be generated when the specified FID is invalid.

3.5.10   GetFID

Read the physical FID – the unique hardware family slot index of the current thread.

UTLEON3 syntax:

```
getfid %reg_rd
```

is defined as a special case of the instruction

```
RDASR %ASR20, %reg_rd
```

with bit i set to '0', and a part of the instruction opcode replaced with a new field op_uT that identifies the instruction getfid; it is equal to b0100.

- %reg_rd – the destination register that will be written with the FID of the current thread.

Instruction format:

op		rd					op3						rs1=20					i	op_uT				unused(zero)								
1	0	x	x	x	x	x	1	0	1	0	0	0	1	0	1	0	0	0	0	1	0	0	0	0	0	0	0	0	0	0	0	0
31	30	29				25	24					19	18				14	13	12		9	8										0

Description:

The instruction returns a unique index of the current thread's family in the hardware family table, i.e. the physical FID (Family Index).

Each hardware family of threads has a unique physical FID in the range 0 to 2**FT_A_WIDTH-2. The FID returned by this instruction in a child thread is the same as the original allocated FID in the parent thread. Physical FIDs can be used for family-global memory management.

Semantics:

- Write the destination register %reg_rd with the index of the current thread's family in the hardware family table.

Complexity:

O(1)

Affected data structures:

- Destination register

Integration in IU3:

- EX – read the FID from the Thread Table.
- WB – write the FID to a register.

Exceptions:

None.

3.5.11 GetTID

Read the physical TID – the unique hardware thread slot index of the current thread.

UTLEON3 syntax:

```
gettid %reg_rd
```

is defined as a special case of the instruction

```
RDASR %ASR20, %reg_rd
```

with bit i set to '0', and a part of the instruction opcode replaced with a new field op_uT that identifies the instruction gettid; it is equal to b0011.

- %reg_rd – the destination register that will be written with the TID of the current thread.

Instruction format:

op	rd	op3	rs1=20	i	op_uT	unused(zero)
1 0	x x x x x	1 0 1 0 0 0	1 0 1 0 0	0	0 0 1 1	0 0 0 0 0 0 0 0 0

```
31 30 29        25 24          19 18        14 13 12     9 8                    0
```

Description:

The instruction returns a unique index of the current thread in the hardware thread table, i.e. the physical TID (Thread Index). Note that the *physical* TID is different from the *logical* thread index that is passed to threads in the register %tl0.

Each thread has a unique physical TID in the range 0 to $2**TT\_A\_WIDTH-2$. Physical TIDs can be used for thread-local stack management. When a thread is created it often needs to obtain a private memory range inside which the thread's software stack, heap and other private data can be stored. This memory area must be exclusive to this thread. One way to determine the location of the private memory area inside any thread is by using its physical TID as an index into a global table of pointers. Alternatively, the target location can be computed in an arithmetic expression from the physical TID.

When a thread terminates and later another unrelated thread is created with the same physical TID, the new thread assumes control of the same memory area. Typically the stack segment of the private memory area is reset (discarded), but the heap segment is kept as there still may be live data objects left over by the previous thread. Depending on the runtime system a local garbage collector may be required for managing the heap segment.

Semantics:

- Write the destination register %reg_rd with the index of the current thread in the hardware thread table.

Complexity:

O(1)

Affected data structures:

- Destination register

Integration in IU3:

- EX – read TID from the Thread Table.
- WB – write the TID to a register.

Exceptions:

None.

3.5.12 Break*

Terminate a family of threads from inside a children thread.

UTLEON3 syntax:

```
break %reg_rs2
break imm9
```

is defined as a special case of the instruction

```
WRASR %reg_rs1, %reg_rs2, %ASR20
```

with bit i set to :

- '0' – the operand is a register
- '1' – the operand is a 9-bit unsigned immediate value

and a part of the instruction opcode replaced with a new field op_uT that identifies the instruction break; it is equal to b1010.

- %reg_rs2 or imm9 – the return family termination code
- 0 means $FID = 0 +$ the type of the stored value ($BREAK$)

Instruction format:

op		rd=20				op3					unused(zero)						i	op_uT			unused(zero)					rs2				
1	0	1	0	1	0	0	1	1	0	0	0	0	0	0	0	0	0	1	0	1	0	0	0	0	0	x	x	x	x	x

31 30 29   25 24   19 18   14 13 12   9 8   5 4   0

op		rd=20				op3					unused(zero)						i	op_uT			imm9									
1	0	1	0	1	0	0	1	1	0	0	0	0	0	0	0	0	0	1	1	0	1	0	x	x	x	x	x	x	x	x	x	x

31 30 29   25 24   19 18   14 13 12   9 8   0

Description:

This instruction terminates a family of threads from within the family based on a dynamic condition. An example use would be a termination of an infinite loop (*step* set to 0). Any thread in a family can terminate the execution of the whole family (the whole execution subtree) it belongs to.

Semantics:

- Terminate creation of new threads in the family.
- Terminate running threads in this family.

- Force cancellation of all synchronization events for the threads in this family (teminate suspended threads that accessed registers marked *PENDING*).
- Write the return value of this family of threads, to be read by the parent thread (the value specified in the `break` instruction).
- Write the termination code *BREAK* to the register specified during the creation of this family (read by the parent thread).

Complexity:

O(n)

Affected data structures:

- Family table – one table entry.
- Thread table.
- Thread scheduler.
- RegFile – write the termination code.
- Processor pipeline.

Integration in IU3:

- RA – read the family termination code.
- EX – execute the instruction (multi-cycle execution ⇒ halt the pipeline).

Exceptions:

None

Note:

Not implemented.

3.5.13   Kill*

Terminate a family of threads from its parent thread.

UTLEON3 syntax:

```
kill %reg_rs1
```

is defined as a special case of the instruction

```
WRASR %reg_rs1, %reg_rs2, %ASR20
```

with bit i set to '0' and a part of the instruction opcode replaced with a new field op_uT that identifies the instruction `kill`; it is equal to b1011.

- `%reg_rs1` – an FID of a family to be killed

Instruction format:

op			rd=20					op3					rs1					i	op_uT				unused(zero)							
1	0	1	0	1	0	0	1	1	0	0	0	0	x	x	x	x	x	0	1	0	1	1	0	0	0	0	0	0	0	0	0

31 30 29   25 24   19 18   14 13 12   9 8   0

Description:

Any thread (from a different family of threads) that knows a correct FID value can force termination (`kill`) of a family of threads identified by the FID value. The FID value is passed through a register.

Semantics:

- Terminate creation of new threads in the family FID.
- Force termination of running threads in this family.
- Force cancellation of all synchronization events for the threads in this family (terminate suspended threads that accessed registers marked *PENDING*).
- Kill all families of threads created by threads in this family (execute `kill`).
- Write the termination code *KILLED* to the register specified during the creation of this family (read by the parent thread).

Complexity:

$O(n)$

Affected data structures:

- Family table – one table entry.
- Thread table.
- Thread scheduler.
- Register file – write the termination code.
- Processor pipeline.

Integration in IU3:

- RA – read a family identifier.
- EX – execute the instruction (multi-cycle execution $\Rightarrow$ halt the pipeline).

Exceptions:

An exception can be generated when an invalid FID (e.g. a nonexistent family of threads) is specified.

Note:

Not implemented.

Chapter 4
The Basic UTLEON3 Architecture

This section summarizes the basic extensions to the LEON3 processor that support the microthreaded model of program execution. Please note that execution of families of threads in hardware accelerators and microthreaded interrupt handling are covered in separate chapters. The structure of the basic UTLEON3 processor is shown in Fig. 4.1.

The major features of the basic UTLEON3 processor are:

- *Legacy* mode for backward binary compatibility with LEON3.
- *Microthreaded* mode for hardware microthreading.
- Automated context switch management through queues of ready and waiting processes, together with the program counter information stored in the register file.
- General-purpose registers used to store thread identifiers for threads that are waiting for register updates from the memory.
- Context switch in zero clock cycles in most cases (three clock cycles in the worst case).
- Register allocation in windows on a per-thread basis.
- Memory access in 16-word burst transactions.
- Non-blocking memory access on cache miss.
- Autonomous register update on D-Cache line fetch.

4.1 Program Execution

The UTLEON3 processor supports execution of both legacy and microthreaded binaries. The execution of the legacy code generates the same results on UT-LEON3 as on LEON3. The UTLEON3 program execution starts in the legacy mode (see Fig. 4.2). After initialization the software switches UTLEON3 to the microthreaded mode using the *launch* instruction. The starting address of the

M. Daněk et al., *UTLEON3: Exploring Fine-Grain Multi-Threading in FPGAs*,
DOI 10.1007/978-1-4614-2410-9_4, © Springer Science+Business Media, LLC 2013

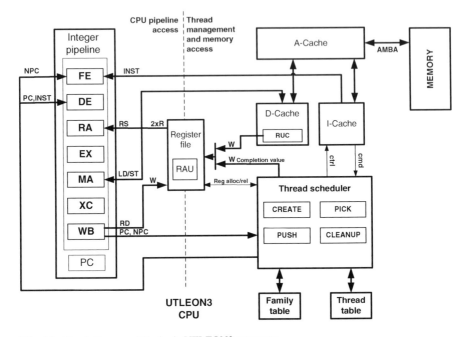

Fig. 4.1 Block diagram of the basic UTLEON3 processor

microthreaded code is passed to the thread scheduler. When the microthreaded execution finishes, i.e. when there are no threads to be executed, the processor switches
back to the legacy mode. This behaviour is very similar to a subprogram call.

A typical UTLEON3 execution consists of these steps:

- Reset – the processor is initialized in the legacy mode.
- launch – the processor is switched to the microthreaded mode.
- Execution of the main (parent) thread in Family 0.
- allocate – allocation of a space in the family table (FT).
- set... – setting the parameters necessary to create a family of threads (FT) in
  the family table.
- create – creation and execution of a family of threads (FT).

  – Execution of the parent thread until it reads a register marked pending or
    waiting.
  – Execution of children threads – each thread can terminate either when it
    finishes its computation, or when the *break* or *kill* instruction is executed.

- Continued execution of the parent thread.
- Processor switches back to the legacy mode when all threads have finished.

The basic unit of microthreaded computation is a family of threads that groups
threads that form one kernel. Each thread can access a set of family global

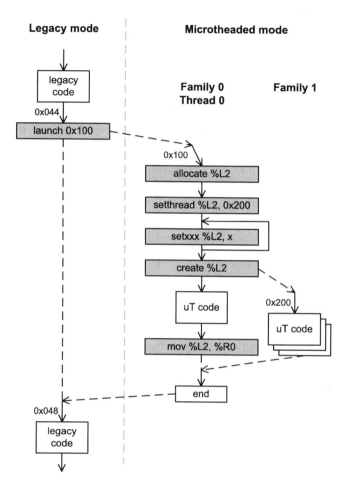

Fig. 4.2 Program flow

registers %*tgi*, thread local registers %*tli*, thread dependent registers %*tdi* and thread shared registers %*tsi*. The global registers are shared among all threads in a family. The local registers are unique for each thread; they serve as a thread local storage. Dependent and shared registers allow passing data from one thread to its direct sibling thread in one direction only – the shared registers of the thread *i* and dependent registers of the thread *i* + *1* are mapped to the same addresses.

All possible bubbles inserted in the processor pipeline due to unsatisfied data dependencies are eliminated by a thread switch. The switch can be either implicit, meaning that it occurs in the runtime when a source register does not contain valid data or on an I-Cache line miss, or it can be requested explicitly by the *swch* modifier. A thread switch causes another thread to enter the processor pipeline, while the current thread returns in the pipeline later when the corresponding data are available.

4.2 Processor Functional Model

Figure 4.3 shows a simplified functional model of the UTLEON3 processor. Several things can be noted. Instructions originate either in the I-Cache or in the scheduler. The I-Cache supplies instructions either to the fetch stage or to the scheduler to implement a zero clock cycle (CC) switch on an I-Cache miss. The program counter is calculated either in the processing pipeline, or a new value is supplied by the scheduler on a context switch. On a D-Cache hit the pipeline operates as the normal SPARC pipeline; on a D-Cache miss (LD) the pipeline marks the destination register as pending and continues processing; the register update happens autonomously in the background; on a D-Cache miss (ST) the value to be written is put in a store queue (not shown).

An overview of the pipeline operation on an instruction or data miss is shown in Figs. 4.4 (legacy mode) and 4.5 (microthreaded mode). The numbers in the boxes between the pipeline stages denote the size of the bubble incurred in the pipeline. In the legacy mode the pipeline stalls on instruction or data misses, while in the microthreaded mode a context switch brings in a new thread that has data ready for execution.

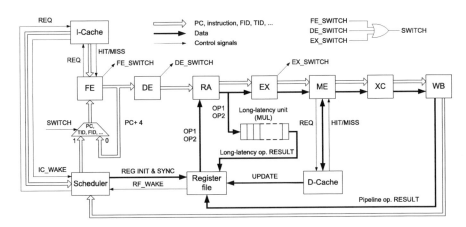

Fig. 4.3 A functional model of the UTLEON3 processor

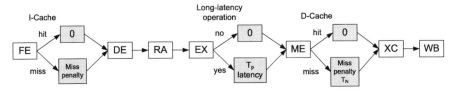

Fig. 4.4 Latencies in the legacy LEON3 processor pipeline

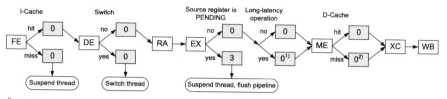

$^{1)}$ When a long-latency operation occurs, the result is marked PENDING and the thread continues.
$^{2)}$ The load result is marked PENDING and the thread continues.

Fig. 4.5 Latencies in the microthreaded UTLEON3 processor pipeline

4.3 Context Switch

The processor implements blocked (coarse) multithreading, meaning a thread is switched out of the pipeline only when an unsatisfied data dependence (i.e. a dependence on a long-latency operation) has been encountered. This improves the single-thread performance, but it requires the pipeline to have fully bypassed stages. A thread switch can be triggered in three pipeline stages:

- *FE_SWITCH* – occurs on an I-Cache miss, the scheduler supplies a new instruction of another thread.
- *DE_SWITCH* – occurs on decoding the *swch* instruction modifier, the scheduler supplies a new instruction of another thread.
- *EX_SWITCH* – occurs on reading a pending register that does not contain valid data, the pipeline is flushed and up to 3CCs are lost.

When a thread switch occurs, the affected thread's instructions in the previous pipeline stages must be flushed. For example, when a thread in the *Execute Stage* of the pipeline reads a register in the *PENDING* state, it must be switched out as the data in the register is not valid, and thus the two previous pipeline stages (*Decode, Register Access*) must be cleared as well – but only if they contain instructions from the same thread as the processor allows for instructions from different threads to be present in distinct pipeline stages at the same time.

To optimize this the architecture allows the programmer to mark instructions where a context switch will occur with the swch modifier (Fig. 4.6); this will cause the thread to voluntarily switch itself out of the pipeline early in the *Fetch Stage*. This is convenient in cases where the compiler or programmer anticipates the instruction depends on a result produced by a long-latency instruction. The voluntary switch causes zero overhead in most cases as there is no pipeline bubble inserted. However, the dependent instruction must be annuled and later re-executed, so in a typical case the synchronization cost is one clock cycle, while without the modifier it is three clock cycles.

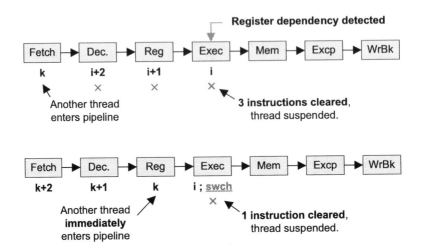

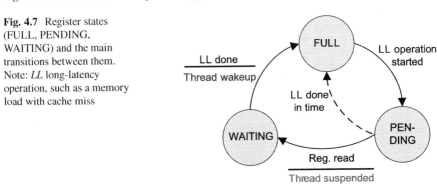

Fig. 4.6 Switch modifier in the processor pipeline

Fig. 4.7 Register states (FULL, PENDING, WAITING) and the main transitions between them. Note: *LL* long-latency operation, such as a memory load with cache miss

4.4 Integer Register File

Fine-grained synchronization and communication between threads in one family, and between the processor pipeline and long-latency units (cache, FPU) is accomplished by a self-synchronizing register file (*i-structures* [1], Fig. 4.7). Each 32b register in the file is extended with a state information. When a register contains valid data, it is in the *FULL* state. If data are to be delivered into the register asynchronously (from the cache, FPU etc.), but they have not been required by the pipeline yet, the register is in the *PENDING* state. Finally, if the data were required by a thread, but were not available at the time, the register is in the *WAITING* state; only on the *PENDING* ⇒ *WAITING* transition the thread has to be switched out and suspended.

Table 4.1 Register structure

State	2b enc	32b value
PENDING	01	N.U.
WAITING	10	TID
FULL	11	DATA

4.4.1   Data Word

Each data word stored in the register file consists of 2 state bits and 32 data bits. A register can be in one of the following states:

- *PENDING* – invalid data. A register is waiting to be updated asynchronously, i.e. independent of the IU3 operation.
- *WAITING* – invalid data. A register is waiting to be updated; furthermore, a thread has tried to read this register when it was *PENDING*. The *thread identifier* (TID) of the thread is stored in the data part.
- *FULL* – data valid.

Table 4.1 shows the coding of the state bits and content of the data part of the register for the register states mentioned above.

4.4.2   Reading and Writing

When reading a register, the integer register file simply reads both the data and state bits. Writing a register is more complicated, since it involves both a read and write operations to set the register state correctly.

When writing a register, the current register state and the register data of the target register have to be read before the register is written. The current and new register states are compared. If the current state is *WAITING* and the new state is *FULL*, a thread *wake-up* request is generated. This request includes *thread identifier* (TID) of the thread that is waiting for the register data.

When the integer pipeline issues a register write, the address of the register is passed through a dedicated R port one clock cycle before the actual write operation. The UTLEON3 integer pipeline write operations are fully pipelined and non-blocking. Write requests that are issued by the units running in parallel with the integer pipeline (register update controller, scheduler, pipelined integer multiplier, trap and interrupt controller, HW families of threads) are served asynchronously through separate R and W ports (one each). As in the case of the integer pipeline, the address of the register is issued on the R port one clock cycle before the actual write operation in order to generate thread wake-up events for registers marked *WAITING*. These write requests are also fully pipelined, but they can be blocked as the port is arbitrated between more components. In this case the blocked component has to wait until the register file finishes the write request issued by a component with a higher priority to avoid data loss.

Table 4.2 Table of possible register state transitions (N.U. = not used, N.C. = not changed)

Instruction %reg_A, %reg_B, %reg_C (add %rs1, %rs2, %rd)					
Current register states			New register state		
%reg_A	%reg_B	%reg_C	%reg_A	%reg_B	%reg_C
PENDING	PENDING	N.U.	WAITING	PENDING	N.C.
	FULL			FULL	
FULL	PENDING	N.U.	FULL	WAITING	N.C.
	FULL			FULL	PENDING
					FULL
Instruction %reg_A, imm, %reg_C (add %rs1, 0x10, %rd)					
Current register states			New register state		
PENDING	imm	N.U.	WAITING		N.C.
FULL			FULL		PENDING
					FULL

4.4.3 State Transitions

Table 4.2 defines valid combinations of register state transitions initiated by the integer pipeline operations.

All the registers are initialized as *FULL* during system reset. Thread global registers are allowed to remain in the *FULL* state only. Any integer pipeline operation that would make them *PENDING* is not permitted, and indicates a program error.

Thread shared registers are set to the *PENDING* state by the hardware thread scheduler only when a thread is being created to enable thread-synchronization based on reading and writing shared and dependent registers. As in the case of thread global registers, any integer pipeline operation that would make them *PENDING* is not permitted, and indicates a program error.

Thread local registers are set to *PENDING* individually when long-latency operations, such as a memory load, are dispatched in the processor pipeline. If a thread local register is in the *PENDING* state, it can be updated only through the asynchronous write port, then its state is set back to *FULL*. A write operation to a *PENDING* thread local register initiated in the integer pipeline indicates a program error.

When a thread reads a register that is *PENDING*, the read event changes the register state to *WAITING*, stores the thread identifier (TID) in the data part of the register, and suspends the thread. Each register can hold only one TID, i.e. only one thread can be waiting for a given register. Note that individual writes of the *PENDING* and *WAITING* states are triggered only in the integer pipeline as a result of operations that read the corresponding registers. The asynchronous write port can set register states only to the *FULL* state through individual register writes.

The *PENDING* to *WAITING* transition is determined in the register access stage of the processor pipeline, but the new state is written in the write back stage. It may happen that at the same time the same register is updated through the asynchronous

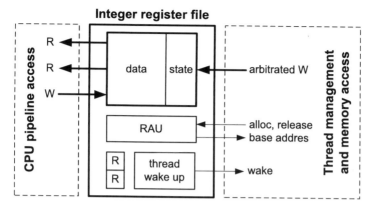

Fig. 4.8 Integer register file schema

port, with its state changed to *FULL*. In this case the request from the processor pipeline must not modify the register, but it must generate a *wake-up* request with the TID taken from the processor pipeline instead of the register.

Accessing a register that is in the *FULL* state will not result in changing its state.

4.4.4 Register File Ports

The register file is a fully synchronous module that has six ports (see Fig. 4.8):

- Two read ports used by the integer pipeline to read the source registers (including their states).
- One read port used by the integer pipeline to read the destination register to get its state and an optional TID for thread wake-up.
- One write port used by the integer pipeline to write the destination register (both the state and data or TID).
- One asynchronous read port for delayed updates and/or thread wake-up.
- One asynchronous write port for delayed updates.

The asynchronous read and write ports (active only in the microthreaded mode) are arbitrated between components that implement namely asynchronous register file update and thread management. These components are listed below together with events that require them to access the register file (starting with the highest priority):

Trap and interrupt controller	– when an interrupt request occurs or an interrupt handler finishes.
Pipelined integer multiplier	– when the result of a multiplication is ready.
D-Cache	– when updating the integer register file with values from the data cache.

Table 4.3 The number of
register file ports

	Read	Write
Integer pipeline access	3	1
Asynchronous access	1	1
Total	**4**	**2**

Scheduler   – when a thread is created.
HW Families   – when a HW family is created or finishes.

When more components attempt to access the register file at once, the component
with the lower priority has to wait. As the register update controller and pipelined
integer multiplier access the register file often, to increase their throughput each uses
a FIFO to store pending write requests.

All operations get dispatched at the execute stage. A single instruction in
the integer pipeline reads up to two source registers. Integer instructions usually
execute in one clock cycle, bypass the memory and get written back to the register
file synchronous with the operation of the integer pipeline. These operations are
scheduled by the compiler.

Memory loads complete synchronously on an L1 cache hit. If a memory access
generates a data cache hit, data is read from the data cache and written to the integer
register file through the integer pipeline write port. If a memory access generates a
miss, the corresponding register is marked as *PENDING*, and a D-Cache line fetch is
requested from the memory subsystem. When the data return from the higher level
memory, they are written to the integer register file through the asynchronous port.

When a register is being written, the current and new register states must be
compared. This means the integer register file needs to perform one additional read
operation prior to each write operation. Two write ports imply two state read ports
used internally in the integer register file. Table 4.3 sums up the integer register file
ports needed.

4.4.5 Register Allocation

The UTLEON3 processor allows each thread to individually specify the number of
required registers from 1 up to the architectural limit of 32 registers per thread given
by the SPARCv8 instruction encoding (Fig. 4.9). This is done through the *.registers*
pseudo-instruction given at the beginning of each thread routine that informs the
scheduler how many registers it must allocate to each thread from the family.

During thread creation the scheduler assigns registers to threads on a per-thread
basis. The allocation is based on assigning a free register window to each thread.
The *Register Allocation Unit* (RAU) is a module that is responsible for allocating
and deallocating regions of registers in the integer register file. The RAU structure
is shown in Fig. 4.10.

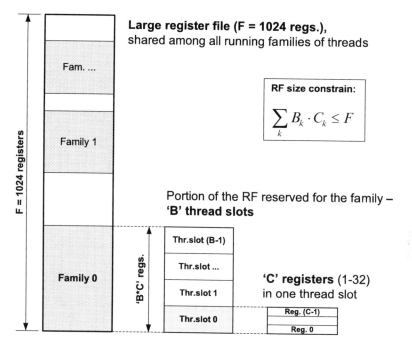

Fig. 4.9 Partitioning of the register file among dynamically executed families of threads

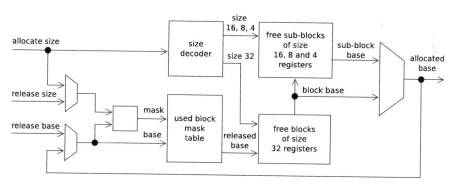

Fig. 4.10 Register Allocation Unit (RAU)

To increase the efficiency of the register file usage the RAU can allocate registers in window sizes 32, 16, 8 and 4 registers. To split a 32-register block to smaller sub-blocks the RAU uses a table of block masks; the table is addressed by the register block base address.

On an allocate request the RAU returns the base address of a free region in the integer register file so that its size is the smallest possible to fit in the number of requested registers. This mechanism allows to minimize the number of unused yet allocated registers in the region. When a request consumes a full block of 32 registers, the *free block* component is used directly.

Requests that can be satisfied with register sub-blocks of sizes 16, 8 and 4 are executed in more steps. First, the RAU checks whether the *free sub-block* component contains a free sub-block of the appropriate size. If so, the sub-block will be used; at the same time if this is the last sub-block in a 32-register block assigned to the *free sub-block* component, the block is marked as used. If such a sub-block is not available, a new block from the *free block* component is assigned to the *free sub-block* component, it will be partitioned to the required sub-block size, and its first sub-block will be used. On allocation the corresponding bits in the used mask are set.

On register deallocation the base address of the register region to be deallocated is required as well as the size of the block to be released. The corresponding bits in the used mask are reset. When all the bits in the used mask of a sub-block are cleared, the block can be released back to the queue of free blocks in the *free block* component.

4.5 Memory Interface

The memory interface in UTLEON3 is derived from LEON3. Memory is accessed through the cache subsystem that consists of an instruction cache and a data cache. Both the *I-Cache* and *D-Cache* share access to the AMBA AHB bus; bus arbitration is done in *acache*.

Memory requests are generated on cache line miss, on cache line eviction and on cache flush. Each request is stored in a FIFO and served on a one-at-a-time basis. UTLEON3 features a write-back data cache with independent store queues for each pending memory request.

When the cache subsystem reaches a situation when all cache lines that can serve a memory request are busy serving previous requests, it suspends the integer pipeline operation until a cache line becomes available.

4.5.1 Data Cache

Memory accesses complete synchronously on an L1 data cache hit. If a memory access generates a data cache hit, data is read from the data cache and written to the integer register file in a subsequent clock cycle. If a memory access generates a miss, the corresponding register is marked *PENDING*, and a cache line fetch is requested from the memory subsystem. When the cache line data return from the higher level memory, they are written later via a dedicated write port to the integer register file. The delayed writes are implemented in the register update controller (RUC).

Possible data cache line fetch scenarios are shown in Fig. 4.11.

The D-Cache line has to be extended to contain information used for delayed register updates (see Fig. 4.12). The two new fields are marked in grey; these contain

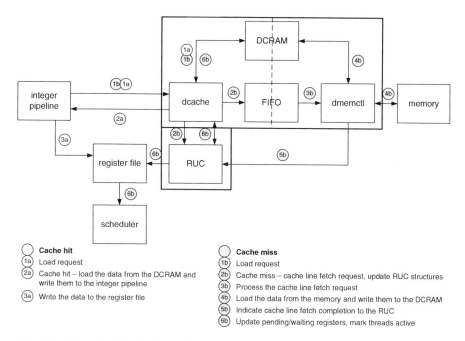

Fig. 4.11 Data cache line fetch scenario

S	Tag	Data	Writeback mask	Head Ptr	Tail Ptr

Fig. 4.12 D-Cache line structure in UTLEON3

the head and tail pointers (physical register numbers) to the linked list structure maintained in the register update controller, and are used to build the update linked lists and trigger register updates.

4.5.2   Register Update Controller

The register update controler (RUC) consists of two finite state machines – one manages a linked list that is used to perform register writes on a data cache line fetch completion, and the other manages a queue of register update requests from the D-Cache on cache line fetch completion. Figure 4.13 shows its block diagram.

One line of the linked list for register writes consists of two parts – a register ID of the next request (*tail*) in the first part with a cache line ID, identified by its (*set, offset, line index*), in the second part. The linked list is implemented using a table.

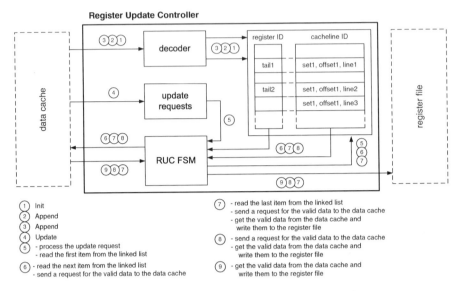

Fig. 4.13 Register update controller (RUC) – block diagram and use scenario

The RUC performs three operations – *init*, *append* and *update*. The meaning of the RUC operations is as follows:

init   – creates the *head* of the list (operation *2b* in Fig. 4.11). The *tail* part of the row addressed by the register ID is set to the current register ID. The address of the required data word within the data cache (set, offset, line index) is stored in the cache line ID part.

append   – appends a new element to the linked list (operation *2b* in Fig. 4.11). The *tail* part of the row addressed by the register ID is not changed. The *tail* part of the row addressed by the previous register ID is set to the current register ID. The address of the required data value within the data cache (set, offset, line index) is stored in the cache line ID part.

update   – when a data cache line fetch finishes, delayed writes to integer registers that are waiting for the cache line get initiated (operation *5b* in Fig. 4.11). The *head* and *tail* values are stored in the update request queue. The RUC update FSM reads the linked list and generates corresponding data cache read requests and integer register file writes.

4.5.3 Instruction Cache

The microthreaded architecture decouples instruction execution from memory operations so that while one group of instructions is being requested or fetched from the memory, another group of instructions (microthread) executes. The instruction

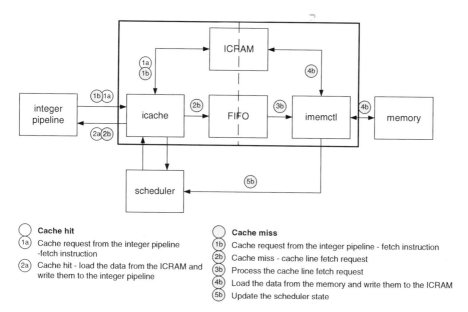

Fig. 4.14 Instruction cache line fetch scenario – requests initiated in the fetch stage

cache serves requests from the integer pipeline and from the thread scheduler; this means it arbitrates the access to the main memory between three sources: the integer pipeline fetch stage, the integer pipeline execute stage, and the thread scheduler.

- The fetch stage access corresponds to the instruction fetch, i.e. to the common CPU execution.
- The execute stage access is generated by the *setthread* instruction that is issued prior to the *create* instruction; this instruction passes the *.registers* word from the memory to the scheduler.
- The scheduler access is required so that the scheduler can check the availability of cache lines required to start threads and mark waiting threads as ready for execution.

The three possible instruction cache line fetch scenarios are shown in Figs. 4.14–4.16.

Microthreading requires that no I-Cache line be evicted if at least one thread exists in the processor (active, waiting or suspended) whose program counter points to an instruction within the I-Cache line; this is to maintain consistency as to which event caused a thread to be suspended (unsatisfied data dependency) or waiting (instruction cache miss), never both at once. Furthermore, on restarting a suspended thread an instruction cache miss must not occur for the restarted instruction. This requires the instruction cache line be extended with a field that contains the number of threads that currently use a given I-Cache line (the grey field denoting the reference counter in Fig. 4.17).

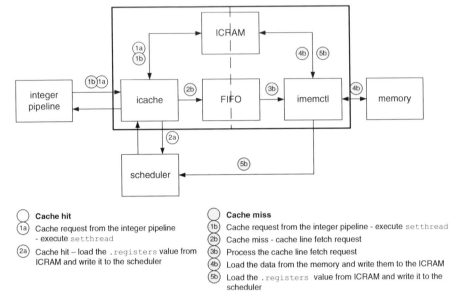

Fig. 4.15 Instruction cache line fetch scenario – request initiated in the execute stage

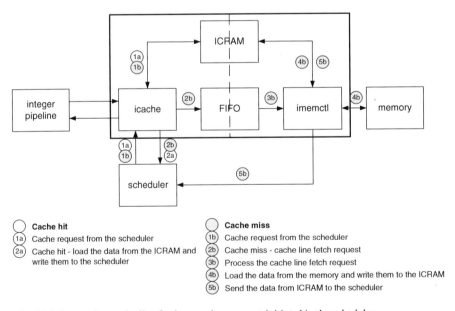

Fig. 4.16 Instruction cache line fetch scenario – request initiated in the scheduler

S	Tag	Data	#References

Fig. 4.17 I-Cache line structure in UTLEON3

4.6 Thread Management

4.6.1 Family Creation

The *allocate* instruction (see Fig. 4.18) gets the first item from a queue of indexes (FIDs) of free family table entries. If the queue is empty, an exception is thrown. The corresponding family table entry is initialized for a new family by default values and updated with a family index (FID) and thread index (TID) of the parent thread. The *FID valid* flag is set and the *family created* flag is cleared as well, which enables the *set...* instructions to modify this family table entry.

The *setstart, setlimit, setstep, setblock* and *setthread* instructions write their operands into the *start, limit, step, block* and *thread/pc* fields of the specified family table entry respectively. The *FID valid* and *created* flags are checked to ensure that parameters are updated only for family entries that have been allocated but not used by the *create* instruction. The *setthread* instruction sets an address of the first instruction of a family code and initiates loading of the corresponding cache-line.

The *create* instruction takes an allocated family entry, and it issues a request to create the family. The *created* flag is set to avoid further writes to this family entry. This step ends the creation process as far as the parent thread is concerned. All other operations happen in hardware in the *thread scheduler* component, asynchronously to the parent thread execution.

The sequence of operations performed by the thread scheduler in order to create a family is as follows:

1. At the time the *create* instruction is being executed, the *thread scheduler* checks whether the family's first cache-line has been already loaded. The scheduler also requires the *.register* word (see Fig. 1.1). The fetch of the cache-line that contains *.registers* has already been initiated by the *setthread* instruction. If the cache-line has been loaded, the family is put in the *create* queue immediately. If the cache-line has not loaded yet, the family is put in this queue on completion of the family's first cache-line fetch. The reference count for this cache-line is not increased since no thread from the family has been created yet.
2. If the *create* queue is not empty and threads of no other families are being created, a family is picked from the head of the *create* queue.
3. The scheduler checks whether the family has been already initialized. If it has been initialized, the processing continues with thread creation. If it has not been initialized, the processing continues with family initialization.

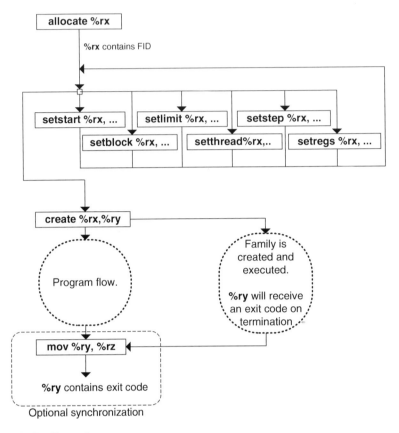

Fig. 4.18 *Family* creation

4. Parameters of the parent thread are loaded in order to calculate the address of the parent global and shared registers for the family being created. These registers are mapped to the local registers of the parent thread.
5. Family initialization is finished, the *family-initialized* flag is set and the process continues with thread creation.

The family creation scenario for one CPU is shown in Fig. 4.19.

4.6.2   Thread Creation

The thread creation process can either allocate new thread table entries and create new threads for a family, or reuse thread table entries that have been allocated to threads of the same family that have already finished their execution and have been cleaned up. The thread table entries are identified by their index (TID).

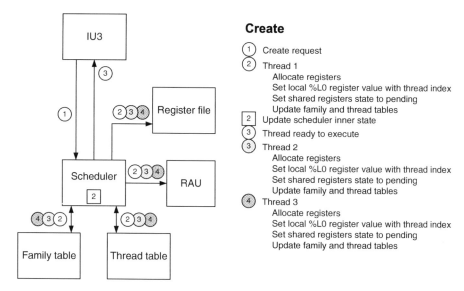

Create

① Create request
② Thread 1
       Allocate registers
       Set local %L0 register value with thread index
       Set shared registers state to pending
       Update family and thread tables
2   Update scheduler inner state
③ Thread ready to execute
③ Thread 2
       Allocate registers
       Set local %L0 register value with thread index
       Set shared registers state to pending
       Update family and thread tables
④ Thread 3
       Allocate registers
       Set local %L0 register value with thread index
       Set shared registers state to pending
       Update family and thread tables

Fig. 4.19 Thread creation scenario

When new thread table entries are allocated, thread creation is started by reading the FID from the *create* queue. If the FID refers to the family that has not been initialized yet, the family initialization phase is performed first. The thread creation process is then performed until:

1. The number of thread table entries that are allocated for the family is equal to the *block size*. This means that this family has reached its maximal allowable number of co-existing threads; in this situation additional threads of the same family can be created only through reusing the thread table entries already allocated by previous threads from this family.
2. All entries in the thread table have been allocated. The FID is put back in the *create* queue, and the creation of new threads is postponed until a thread table entry is released.
3. A TID from the family occurs in the *cleanup* queue. Reuse of thread table entries has a higher priority than allocating new thread table entries, thus the FID is put back in the *create* queue, and the TID from the *cleanup* queue is processed.
4. All family threads have been created.

When reusing thread table entries, thread creation is started by reading the TID from the *cleanup* queue. The TID can be reused only for threads from the same family. When all threads from a family have been created, the thread table entry is released to the pool of free thread table entries, and can be used by any family.

Note that the cleanup process has a higher priority, so that the next thread in the index sequence will reuse the old thread table entry.

Thread creation is performed by the hardware thread scheduler and consists of these steps:

1. When using a new thread table entry (as opposed to reuse of an already allocated thread table entry for a TID that has been cleaned up), allocate a new thread table entry, evaluate offsets in the register file for the thread local and dependent registers, and include the thread in the family member list.
2. Initialize values and states of the registers: the %tl0 register contains the thread index, all shared registers are set to *PENDING*.
3. Request the thread first instruction from the instruction cache. This event increases the reference counter in the corresponding cache-line.
4. Update the thread table entry; update the family table entry.
5. If the instruction requested in Step 3 has been provided in the meantime, the thread is put in the *active* queue. Otherwise the thread is marked as waiting, and will be put in the *active* queue later when the cache finishes the corresponding cache-line fetch.

4.6.3   Instruction Fetch

Instructions that are to be issued to the UTLEON3 processor pipeline come from two sources:

- The *instruction cache* provides instructions during a regular program run in the same way as in the original LEON3 processor.
- The *thread scheduler* provides the first instruction that the thread should start with after a thread switch.

The thread scheduler provides all the necessary data required to initiate thread execution to the processor pipeline inputs. The data of the first active thread are held at these inputs until a thread switch occurs. When a switch occurs, the data of the next active thread are provided, and the thread enters the pipeline with zero clock cycle latency. If there are no active threads available, the *force-nop* interface signal is set, meaning that the fetch stage should issue NOP instructions to the processor pipeline until a thread becomes active.

4.6.4   Thread Switch

A thread switch (context switch) occurs in the following situations:

- On an *I-Cache miss* – this event occurs in the *fetch stage* of the processor pipeline when a *cache miss* is detected at the I-Cache interface. The current thread becomes *WAITING*, and is expected to be woken up by I-Cache at the time the corresponding cacheline load will have finished.

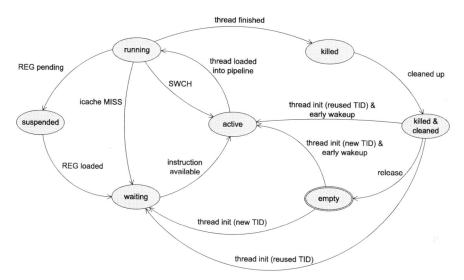

Fig. 4.20 Thread states

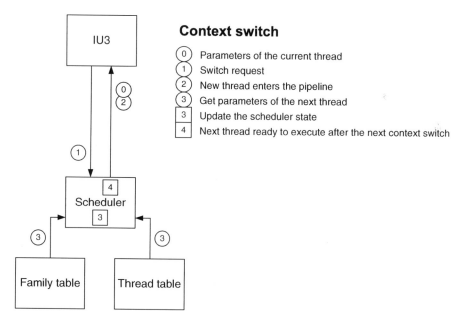

Fig. 4.21 Context switch scenario

- On an explicit *thread switch* request – this event occurs in the *decode stage* of the processor pipeline when the *SWCH* instruction modifier is detected. The current thread leaves the pipeline, nevertheless it remains *ACTIVE*. This switch occurs even if there are no other active threads available.

- On accessing a *pending register* – this event is invoked in the *execute stage* of the processor pipeline when at least one instruction operand is in the *PENDING* state. The current thread becomes *SUSPENDED*, and is expected to be woken up by the register file at the time the corresponding register is updated and set to the *FULL* state.

Switches due to an I-Cache miss and thread switch request do not introduce any bubbles in the processor pipeline. On the other hand, switches due to operations with pending registers introduce up to three clock-cycle long bubbles since the decode stage, register-access stage and execute stage of the processor pipeline have to be annuled (if the TID of the instructions in those stages is the same as of the instruction that caused the thread switch).

Thread switch does NOT occur in the following situations:

- Execution of a long-latency operation – the operation marks the destination register as *PENDING*, but does not force thread switch. A switch occurs only at the time when the *PENDING* register is accessed by a subsequent instruction.
- Crossing of an I-Cache line boundary – this does not cause thread switch provided the target cacheline is present in the instruction cache, i.e. a cacheline request hit.

The thread switch scenario is shown in Fig. 4.21, while Fig. 4.20 shows the thread state diagram.

Chapter 5
UTLEON3 Programming by Example

The purpose of this chapter is to explain basic rules of a practical microthreaded assembler-level programming for UTLEON3. This will be achieved by explaining a couple of increasingly complex assembler programs. The first program, called VADD, will illustrate the basic structure of an assembler microthreaded program. The second program, called FIB, will show how to use *family shared registers* to implement loop-carried variables. Finally, the third program, called CCRT, will demonstrate a *continuation create*.

5.1 Programming Framework

Each program displayed here constitutes a self-contained code which can be executed directly on the UTLEON3 processor. To achieve that, common hardware initialization routines are provided in the programming environment. These routines execute in the legacy mode and then the control transfers to the microthreaded mode, as depicted in Fig. 4.2. In all the examples herein we assume the launch instruction has been just executed, starting our code at the ut_main label.

5.2 Program VADD: Integer Vector Addition

We will demonstrate the basics of microthreaded programming on the most simple program imaginable: integer vector addition. The algorithm of the program is displayed in Listing 1. This algorithm consists of one *for-loop*. The looping construct at line 1 specifies the number of iterations, while its body at line 2 describes the algorithm to take on each iteration. Assuming the input and output arrays are distinct we can directly see the loop is completely data-parallel and thus eligible for microthreaded implementation.

M. Daněk et al., *UTLEON3: Exploring Fine-Grain Multi-Threading in FPGAs*, DOI 10.1007/978-1-4614-2410-9_5, © Springer Science+Business Media, LLC 2013

Listing 1 : Program VADD – the algorithm

```
   for ( int  i = 0;  i <= VECLEN−1; ++i )
2        wdata0[i] = rdata0[i] + rdata1[i];
```

Listing 2 : Program VADD (mtsparc assembler listing)

```
   /* exported  assembler  symbols */
2  .global ut_main , ut_vadd , rdata0 , rdata1 , wdata0

4  /* constants */
   .equ     CACHELINE , 16*4
6  .equ     BLOCKSIZE_1 , 4
   .equ     VECLEN, 256        /* array size */
8
   .section " . bss "           /* uninitialized DATA in RAM */
10 .align 16

12 /* allocate space in the bss data segment */
   .lcomm    rdata0 , (VECLEN*4)                /* input array 0 */
14 .lcomm    rdata1 , (VECLEN*4)                /* input array 1 */
   .lcomm    wdata0 , (VECLEN*4)                /* output array 0 */
16
   /* ========================================================= */
18 . text                                      /* program text in ROM */
   . align CACHELINE
20        .registers 0 0 31   0 0 0             /* 0 GR, 0 SR, 31 LR */
   ut_main :
22        set  rdata0 , %tl0                    /* sub−%tg0 = rdata0 */
          set  rdata1 , %tl1                    /* sub−%tg1 = rdata1 */
24        set  wdata0 , %tl2                    /* sub−%tg2 = wdata0 */

26        /* the family */
          allocate %tl20
28        set (VECLEN−1)*4, %tl21
          setlimit %tl20 , %tl21                /* set index upper value */
30        setstep %tl20 , 4                     /* set index increment */
          set ut_vadd , %tl21
32        setthread %tl20 , %tl21               /* set thread text address */
          setblock %tl20 , BLOCKSIZE_1
34                                              /* set the blocksize parameter */
          create %tl20 , %tl20
36
          mov %tl20 , %tl21 ; SWCH !
38                                              /* wait for family to terminate */
          nop ; END !
40 . size ut_main , .−ut_main
```

The corresponding microthreaded assembler source code of the algorithm is given in Listings 2 and 3. The source code can be divided into three parts:

Listing 3 : Program VADD (mtsparc assembler listing) cont'd

```
   /*  =========================================================  */
42 /*  vector  add  */
   /*  %tl0  =  index  'i'
44 *  %tg0  =  rdata0 ,  %tg1  =  rdata1 ,  %tg2  =  wdata0
   */
46 .align  CACHELINE
         .registers  3 0 4   0 0 0              /*  3  GR,  0  SR,  4  LR  */
48 ut_vadd :
         ld  [%tg0  +  %tl0 ] ,  %tl1
50       ld  [%tg1  +  %tl0 ] ,  %tl2          ;  SWCH  !
         add %tl1 ,  %tl2 ,  %tl3
52       st  %tl3 ,  [%tg2  +  %tl0 ]          ;  END   !
   .size  ut_fam1 ,  .−ut_fam1
```

1. Lines 1–16: assembler headers, data space reservations.
2. Lines 17–40: family creation, corresponding to line 1 in Listing 1 (looping construct).
3. Lines 41–52: the thread definition, corresponding to line 2 in Listing 1 (loop body).

The assembler source code will be explained in a bottom-up fashion. The *for-loop* body at line 2 in Listing 1 is implemented as thread at lines 41–52 in Listings 2 and 3. Threads will be run for each value of the index variable i. From the loop-body 'wdata0[i] = rdata0[i] + rdata1[i];' we see that four symbols are required: rdata0, rdata1, wdata0, and i. The first three symbols are constant within the body. We can either load their values each time the iteration is taken, or use *global* registers to hold them. The code presented uses three thread global registers %tg0, %tg1, and %tg2 to hold the base addresses of arrays rdata0, rdata1, wdata0.

The number of thread global registers used has to be declared in the .registers directive at line 47. The .registers pseudo-instruction inserts a specially formatted data word (*Register Info*) into the program text informing the UTLEON3 processor how many local registers to allocate to each thread and how many global and shared registers are used. This information uniquely defines the mapping between an architectural register number (encoded in the 5-bit fields of the SPARC instructions) and the physical register address within a large processor register file.

The organization of the thread prologue code at lines 46–48 is very important as it has to adhere to the layout presented in Fig. 1.1. First, the code has to be aligned on a cacheline beginning (line 46), i.e. a multiple of 16 words. In the very first word of every instruction cacheline (offset 0) there is the *Control Word* which is inserted automatically by the assembler tool and needs not be inserted explicitly by the assembler programmer. Then, at offset 1, the *register information* word comes. This word is specified at the beginning of the thread (only on its first cacheline) by the assembler programmer through the use of the .registers pseudo-instruction. In the example program we specify three thread global registers, 0 thread shared registers, and four thread local registers.

Each thread has a set of thread local registers, referenced as %tl0–%tl$n$ in the assembler source. The actual number of these registers is specified in the .registers pseudo-instruction. Note that the total number of thread-visible registers has to be less than or equal to 31, i.e.

$$GR + 2 \cdot SR + LR \leq 31.$$

In our case we have $3 + 2 \cdot 0 + 4 = 7 \leq 31$.

The thread local register %tl0 is special as it automatically receives the iteration index when the thread is started. In the example code we use the iteration index in %tl0 to offset the array base addresses in the three global registers. To be able to do so the index has to be scaled by 4 (the machine word size) so that the address obtained is correct. This is easily achieved by scaling the loop limit and iteration step during family creation at lines 28 and 30. Finally, to end a thread the END modifier is specified with the st instruction at line 51.

Moving further up, the second part of Listings 2 and 3 (lines 17–40) contains mainly the thread family initialization code. This part is itself a thread, and so it has to use a similar prologue code (at lines 19–21) as the one discussed previously. Because the ut_main thread is started by the launch instruction, its register layout is fixed and thus the corresponding .registers pseudo-instruction has to specify the number of registers exactly as shown.

A new family is allocated at line 27, and its handle, called *Family ID* (FID), is stored in register %tl20. Then a couple of the set... instructions (setstart, setlimit, setstep, setthread, setblock) at lines 28–33 are executed on this handle to setup the family. The order of these instructions does not matter. The setstart instruction is not used here as the starting index value automatically defaults to zero. The setlimit instruction (line 29) is used to set the last valid value of the index counter. This is VECLEN-1 in our case. To enable the use of the index directly as array offset we perform an optimization and scale it by 4. Consequently, the index increment value, set by the setstep instruction at line 30, is also 4 instead of 1 which is specified in the original algorithm. The setthread instruction has to be always present to set the thread text code address. Finally, the setblock instruction can be used to specify the family *blocksize* parameter.

Before the create instruction is issued at line 34, the family global (and shared) registers have to be initialized. This is done in advance at lines 22–24. The values are written into local registers of the parent thread, starting at register %tl0.[1] They will be made visible in the child family of threads as registers %tg0–%tg2. Note that the architecture requires that these registers be in the *FULL* state upon family creation.

---

[1] The setregg/setregs instructions, currently not implemented, could be used to specify register offset where the register mapping would be based.

The child family of threads is started by the `create` instruction at line 34. The first argument specifies the FID, while the second one specifies the destination register for the family exit code–the so-called *sync register*. Upon creation the sync register is set to *PENDING*, and the parent thread continues running. Usually we want to make the parent thread wait until the child family has completed. This is achieved by reading the sync register at line 36 by a simple `mov` instruction. As we expect the `mov` instruction to block due to accessing the *PENDING* source register (`%tl20`) the instruction is annotated with `SWCH`, so the parent thread is switched out of the pipeline prematurely.

Once the child family completes, the sync register receives the family exit code, and the parent thread is woken up at line 36. The `ut_main` thread is ended at line 37 on execution of the instruction with the `END` modifier. As this thread was created from the launch instruction, its completion causes the processor to switch back to the legacy mode.

5.3 Program FIB: Fibonacci Number Generator

The second program to be shown calculates the Fibonacci sequence (1, 1, 2, 3, 5, 8, 13, 21, ...), and is called `FIB`. This example will demonstrate the use of shared registers to implement loop-carried variables. Loop-carried variables are those whose values are calculated in one loop iteration and then consumed in another, dependent one. As they represent true data dependencies, they cause the loop to be executed sequentially (in general it is better to avoid them if possible).

Listing 4 shows the algorithm we will use to implement the Fibonacci generator. Each Fibonacci number in the sequence is simply a sum of the two preceding numbers in the series. This computation can be seen at line 8 in the listing, where the local variable '**int** fn' is calculated. Each newly calculated number in the series is stored in an output array '**int** fibdata [FIBNUM+2]'. The two initial numbers of the Fibonacci series are '1', and they are written at line 5.

The most interesting part of the algorithm is the use of the two variables '**int** p1, p2' to hold the two last calculated values in the series. Alternatively, the two predecessor values could be obtained directly from the 'fibdata' array, but this approach would require more `load` instructions, and–perhaps more importantly–it would require a totally sequential implementation of the loop.

Listings 5 and 6 displays the microthreaded implementation of the algorithm. The loop body is implemented as a thread at lines 40–52. The thread has one global register, two shared registers, and two local registers. The global register %tg0 contains the output array base address ('fibdata'). This address is set into the register just before family creation at line 21. At line 24 the address is advanced by 2 integer-sizes so that it points to ' fibdata [2] '. This simplifies pointer arithmetic in the thread routine.

The two shared registers are actually visible in the thread routine twice: first as the so-called *dependent registers* (%td0, %td1) which hold values from the immediately preceding thread, and second as the *shared registers* (%ts0, %ts1)

Listing 4 : Program `FIB` – the algorithm

```
   int fibdata[FIBNUM+2];                        /* the output array */
2  int p1, p2;                                /* previous two fib. numbers */

4  /* start with ones */
   p1 = p2 = fibdata[0] = fibdata[1] = 1;
6
   for (int i = 0; i <= FIBNUM-1; ++i) {
8      int fn = p1 + p2;                           /* new f-number */
       p1 = p2;
10     p2 = fn;
       fibdata[i+2] = fn;
12 }
```

---

to hold this thread output values to be passed to its successor thread. The loop-carried variable '**int** p1' (that contains the last but one preceding Fibonacci number) is mapped to the %td0 (input) and %ts0 (output) registers. Similarly, the variable '**int** p2' is mapped to the %td1 (input) and %ts1 (output) registers. The initial values of the input dependent registers (%td0, %td1) of the first thread in the family are set to one at lines 18–19. Notice how the parent's local registers %tl0, %tl1, and %tl2 are mapped to the thread registers %tg0, %td0, and %td1.

The thread output shared registers (%ts0, %ts1) are always initialized to the *PENDING* state. Therefore, when a dependent thread $i + 1$ tries to read its input (line 48, %td0 and %td1), it is in fact reading the thread i output (%ts0, %ts1), and as the values have not been produced yet, the thread $i + 1$ is blocked and must wait. This way the correct thread ordering is established even when individual threads are created and executed concurrently. However, once a value is written to a thread output shared register, it must not be changed. Apart from the explicit synchronization through the shared registers, all threads in a given family are (conceptually) created and executed concurrently. In the `FIB` example in all threads the st instructions at line 51 are independent of each other and thus executed concurrently.

5.4 Program CCRT: Continuation Create

The program CCRT illustrates a so-called *continuation create*. 'Continuation create' is a microthreaded construct to emulate a Unix-like *fork()* command. It allows a parent thread to create a detached child family that will run and terminate independently of the parent. The child family can terminate before or after the parent has terminated.

Listing 5 : Program `FIB` (mtsparc assembler listing)

```
   /* CONSTANTS */
 2 .equ     FIBNUM, 16
                 /* how many numbers in the sequence to calculate */
 4 .equ     BLOCKSIZE, 4
   .equ     CACHELINE, 16*4
 6 .global  ut_main

 8 .section ".bss"
   .align 16
10 .lcomm   fibdata, ((FIBNUM+2)*4)
                 /* calculated fib. series is stored here */
12
   /* ============================================================ */
14 .section ".text"
   .align CACHELINE
16         .registers 0 0 31   0 0 0          /* 0 GR, 0 SR, 31 LR */
   ut_main:
18         /* set the first two Fibonacci numbers for
           * the first thread, i.e. the thread 0 dependencies
20         * registers */
           mov 1, %tl1                              /* sub-%td0 */
22         mov 1, %tl2                              /* sub-%td1 */
                 /* store the first two Fibonacci numbers: 1, 1 */
24         set fibdata, %tl0                        /* sub-%tg0 */
           st %tl1, [%tl0]
26         st %tl1, [%tl0+4]
           add %tl0, 8, %tl0                  /* %tl0 = @fibdata[2]; */
28
           allocate %tl20
30         setstart %tl20, 0                         /* start := 0 */
           set (FIBNUM*4-1), %tl22
32         setlimit %tl20, %tl22               /* limit := fibnum */
           setstep %tl20, 4                    /* step := 4 */
34         setblock %tl20, BLOCKSIZE
           set fib_thr, %tl21
36         setthread %tl20, %tl21
           create %tl20, %tl20
38                 /* wait for family to terminate */
           mov %tl20, %tl21
40         nop ; END
   .size   ut_main, .-ut_main
```

The main attributes of a continuation create are:

- The parent does not wait for a child family to complete. Therefore it specifies %r0 as the exit code synchronization register.
- The child family does not have any global or shared registers. This is because these registers are allocated in the parent thread and they are only made visible

---

Listing 6 : Program FIB (mtsparc assembler listing) cont'd

```
    /*  ==========================================================  */
44  .align CACHELINE
              .registers 1 2 2   0 0 0              ! 1 GR, 2 SR, 2 LR
46              /* %tl0 = result word offset [thread index] */
                /* %tg0 = address of fibdata */
48              /* %td0,%ts0 = second-to-last fibonacci number
                 *             [i-2], p1 */
50              /* %td1,%ts1 = last fibonacci number
                 *             [i-1], p2 */
52  fib_thr:
              add %td0 , %td1 , %tl1              /* fn = p1 + p2; */
54            mov %td1 , %ts0                     /* p1 = p2; */
              mov %tl1 , %ts1                     /* p2 = fn; */
56            st %tl1 , [%tl0+%tg0]  ; END !
    .size     fib_thr , .-fib_thr
```

---

in the child (mapped to its register space). But if the child is expected to outlive its parent, once the parent is gone, the mapping would be corrupt. If parameters are to be transported to the child, it must be done through global variables, or through the family index counter.

The continuation create will be demonstrated on an artificial algorithm in Listing 7. The algorithm is written using an ad-hoc *xfork()* function. It has similar meaning as the well-known Unix *fork()* function call, only the *xfork()* does not separate memory spaces of the parent and child, i.e. it forks a thread, not a process.

The algorithm in Listing 7 creates a one-thread child family at line 9. Then the parent thread enters the busy-waiting loop at line 24, where it awaits completion of the other part of the algorithm. The completion is signalled through a classical global memory–variable 'resv[1]'. Meanwhile the child thread at line 11, running in parallel with the parent, checks the value of another global variable 'resv[0]'. If its value is greater than zero, the thread decrements it (line 14) and then forks another thread–this is the second continuation create in this example (lines 15–17). Otherwise, if the value of 'resv[0]' is zero, it signals (at line 20) to the original parent thread (waiting at line 24) that the first half of the algorithm has finished. Notice that in general this signalling leaf thread is *not* the first child thread created by the first parent thread.

A microthreaded implementation of the algorithm is displayed in Listings 8 and 9. First of all the child family *must not* have any global or shared registers (line 35). This is because these registers are allocated in the parent thread, and they are only made visible in the child (mapped to its register space). But if the child is expected to outlive its parent, once the parent is gone, the mapping would be corrupt.

Listing 7 : Program CCRT – the algorithm

```
   const int CCLENGTH = 10;
 2 int resv[3];

 4 void main()
   {
 6     resv[0] = CCLENGTH;
       resv[1] = resv[2] = 0;
 8
       if (xfork() == 0) {
10                                              /* I am the child */
         child:
12         if (resv[0] > 0) {
                                   /* continue the continuation */
14             resv[0] = resv[0] - 1;
               if (xfork() == 0)
16                                   /* another child created */
                 goto child;
18         } else {
         /* write stop code to signal the original grand-* parent */
20             resv[1] = 1;
           }
22     } else {
       /* I am the parent; periodically check contents of 'resv[1]'*/
24         while (resv[1] == 0)
               ;
26         resv[2] = 1;
       }
28 }
```

There are two create instructions used in the 'continuation' sense (lines 21 and 47) in the example. They can be easily recognized for they specify the **%r0** register as the synchronization register where the family exit code is to be eventually written. It must be done this way; if a common local register was used, the register could be written by the hardware well after the parent terminated, which would corrupt the processor state for another unrelated thread to which the register was allocated in the meantime.

The SWCH modifier used with the ld instruction at line 24 is very important. Lines 23–27 implement the busy-waiting loop of the parent thread. Without the SWCH annotation three unnecessary bubbles would be inserted in the processor pipeline due to the pending state of the synchronization register accessed by the next instruction.

Figure 5.1 shows the execution profile of the CCRT program. In the figure the time is advancing from the left to right. The horizontal lines represent various threads that run in the processor. The first row corresponds to the ut_main thread.

Listing 8 : Program CCRT (mtsparc assembler listing)

```
     /*  CONSTANTS  */
 2   .equ      CCLENGTH,  10              /*  number  of  continuation  creates  */
     .global  ut_main

 4
     .section  ".bss"
 6   .align  16
     .lcomm    resv,  (3*4)              /*  allocate  the  output  resv  array  */

 8
     /*  ============================================================  */
10   .section  ".text"
     .align  CACHELINE
12           .registers  0  0  31  0  0  0       /*  0  GR,  0  SR,  31  LR  */
     ut_main:
14           set CCLENGTH,  %tl1
             set resv,  %tl2
16           st %tl1,  [%tl2]                    /*  resv[0]  =  CCLENGTH;  */
             st %r0,  [%tl2+4]                   /*  resv[1]  =  0;  */
18           set child_thr,  %tl21
             allocate  %tl20
20           setthread  %tl20,  %tl21
             create  %tl20,  %r0                 /*  'cont−create'  */

22
     1:                  /*  periodically  check  contents  of  'resv[1]'  */
24           ld  [%tl2+4],  %tl1  ;  SWCH
             cmp %tl1,  1
26           bne  1b
             nop

28
             set  1,  %tl1
30           st %tl1,  [%tl2+8]  ;  END
     .size    ut_main,  .−ut_main
```

Fig. 5.1 Execution profile of the CCRT program when executed on the processor

---

After the initialization the first child thread is created. Its life span is displayed in
the second row. From this child several other new threads are created progressively
(the other rows), while the old ones terminate–this is the continuation create at line
47 of Listings 8 and 9. In the meantime the main thread executes the busy-waiting

Listing 9 : Program CCRT (mtsparc assembler listing) cont'd

```
     /* ============================================================ */
34  .align CACHELINE
          .registers 0 0 8 0 0 0              /* 0 GR, 0 SR, 8 LR */
36  child_thr:
          set resv, %tl1
38        ld [%tl1], %tl2
          subcc %tl2, 1, %tl2
40        bneg 2f          /* if negative, stop the continuation */
          nop
42
          st %tl2, [%tl1]                     /* resv[0] = resv[0] − 1; */
44        set child_thr, %tl3
          allocate %tl4
46        setthread %tl4, %tl3
          create %tl4, %r0 ; END
48                      /* 'cont−create' and end this child */

50  2:    /* write stop code to signal the waiting ut_main thread */
          set 1, %tl2
52        st %tl2, [%tl1+4] ; END
    .size   child_thr, .−child_thr
```

loop if it has a chance to do so (the first row). Finally, once the last thread (leaf),
displayed in the 12th row, signals its end to the main thread by writing 'resv[1]', the
main thread can exit the busy-waiting loop and finish.

Part II
Implementation

Chapter 6
UTLEON3 Implementation Details

This chapter describes the key functions and implementation of the key new components in UTLEON3. An analysis of performance data and used resources for the Xilinx Virtex 5 LXT xc5vlx110t-1-ff1136 FPGA as well as tcbn90ghptc TSMC 90 nm technology can be found in Appendix E.

6.1 Integer Register File

The integer register file consists of memory blocks (32b data, 2b register state), *thread wake up* request generator, data multiplexer used for arbitration of an asynchronous access port between the data cache, thread scheduler or integer multiplier, register state cleanup logic and register allocation unit. A block diagram of the integer register file is shown in Fig. 6.1.

The memory blocks, both the 32b data and 2b register state parts, are built from dual-port memories (BRAM). As the integer register file needs six ports (two write and four read ports) instead of the two ports provided by the basic BRAM entity in an FPGA, the number of ports is increased by time division multiplexing. The integer register file works with a clock frequency that is three times faster than the main design clock frequency (see Fig. 6.2).

6.1.1 Register Allocation Unit

The register allocation unit (RAU) is responsible for allocating and releasing blocks of registers. The RAU structure is shown in Fig. 6.3. The allocate and release requests are generated in the thread scheduler where register blocks are assigned to threads. The allocated register blocks can be of size 32, 16, 8 and 4 registers, sub-blocks of sizes 16, 8 and 4 are partitioned from blocks of size 32. Blocks of size

M. Daněk et al., *UTLEON3: Exploring Fine-Grain Multi-Threading in FPGAs*,
DOI 10.1007/978-1-4614-2410-9_6, © Springer Science+Business Media, LLC 2013

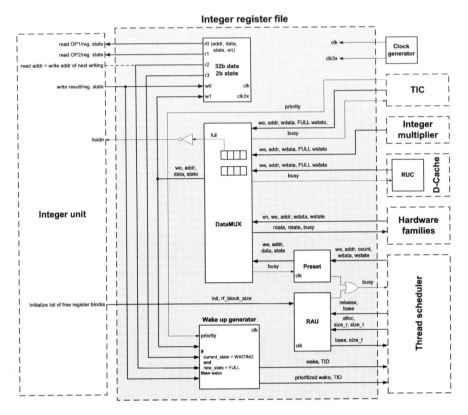

Fig. 6.1 Integer register file – block diagram

32 are stored in the *FREE BLOCKS* FIFO structure. Blocks of sizes 16, 8 and 4 use a more complex data structure called *PARTIALLY ALLOCATED BLOCKS* (PAB). The PAB uses a valid bit, a base address and an offset for each valid block size. Each block of size 32 has its bit mask of used sub-blocks, the masks are stored in an extra table addressed by the block base address.

On an *allocate* request the RAU checks whether the PAB contains a record for the requested block size (its valid bit is active). If so, the next part of the PAB block will be used. If the allocation uses the last free sub-block in a block, its valid bit will be reset. If not so, a new block will be taken from the FIFO, its address will be written to the PAB, it will be partitioned to sub-blocks of the requested size, and the first sub-block will be allocated. On sub-block allocation the corresponding bits of the used mask are set in the mask table.

On sub-block release the corresponding bits of the used mask are reset in the mask table. When all the bits are cleared for the current block (*IS_EMPTY* component in Fig. 6.3), the block is returned back to the FIFO.

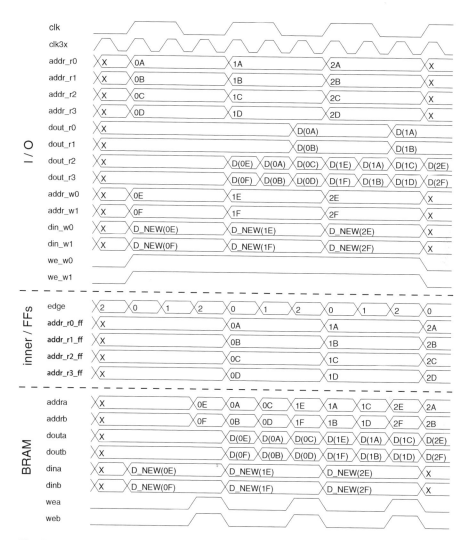

Fig. 6.2 Integer register file – time division multiplexing

6.1.2 Register Value and State Preset Logic

Register value and state preset logic (PRESET) is a simple write request generator. It receives requests from the thread scheduler for initializing continuous blocks of registers with the same value and state. The request comprises the address of the first register and the number of registers to be written. Addresses of the next registers are computed from the first one by incrementing them. The busy signal is kept active until all registers are written.

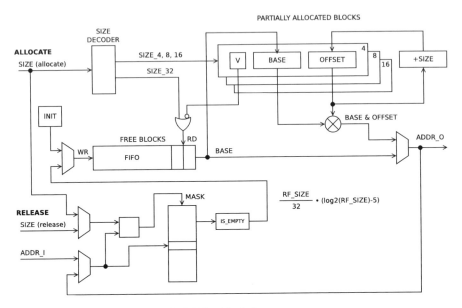

Fig. 6.3 The register allocation unit (RAU) – block diagram

6.1.3 Wake Up Generator

The wake up generator (see Fig. 6.4) issues requests to the thread scheduler to wake up threads on completion of relevant register updates. This means that the wake up generator is active with every write request to the register file. To generate a wake up request the unit needs to know the current register state (all writes to the register file are preceded by a read operation), the new register state (the state that will be written to the register) and the current TID to identify the thread to be woken up. If the current state is *WAITING* and the new state is *FULL*, the generator issues a wake up request.

The generator distinguishes between write operations coming from the processor pipeline (IU3) and write operations on the shared write port (data cache, scheduler, multiplier, ...). As the wake up requests can arise at the same moment, the generator has to manage the requests by their priority (IU3 requests have the higher priority). To avoid losing requests from the shared port the generator may use a FIFO to store them.

A wake up request from the shared port can be supplemented with a high-priority flag for requests coming solely from the trap and interrupt controller (see Sect. 9). It causes prioritized wake up request to the thread scheduler, which can wake up the corresponding thread preferably.

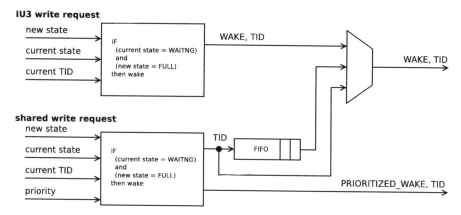

Fig. 6.4 The wake up generator – block diagram

6.1.4 Data Multiplexer

The data multiplexer (DataMUX) is responsible for arbitrating access to the asynchronous ports (one R and one W) between all components that update the register file asynchronously (register update ctl, scheduler, etc.). It accepts all read and write requests, and consequently it controls the read-write sequence to generate wake-up requests properly. Optionally the DataMUX inputs can be extended with FIFOs to store pending requests in the case of multiple accesses. In the current implementation FIFOs are used for the register update controller and the integer multiplier. If two components access the asynchronous port at the same time, the DataMUX halts the component with the lower priority or puts its request in the FIFO if the FIFO is not full. As the integer multiplier operates synchronously with the integer pipeline, it cannot be halted separately, but the whole system must be halted using the *holdn* signal.

6.2 Family and Thread Tables

6.2.1 Family Table

The family table consists of a data memory and an allocation unit. The family table unit block diagram is shown in Fig. 6.5.

The data memory holds global parameters and states of families (see Table 6.1). Single fields of the family table are merged into groups according to the number and type of components that access these fields in order to reduce the number of required memory ports and utilize wide FPGA BlockRAMs efficiently. Each group is mapped to a dual-port memory soft macro (DP RAM) that is implemented using one or more dual-port BRAMs.

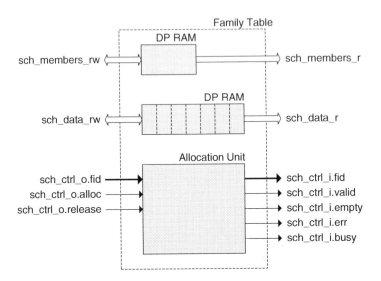

Fig. 6.5 Family table – block diagram

The allocation unit manages the list of family table entries. It consists of a queue of unused entries (their indexes, FIDs) in the family table and a 1-bit wide RAM with a flag that indicates which indexes have been allocated and thus can be accessed by read/write operations.

Operations of the allocation unit:

- *Allocate* returns the index of the first unused family table entry and removes this index from the queue of unused FIDs.
- *Release* puts the index back to the queue of unused FIDs.
- *Check* checks whether read/write operations access valid entries (i.e. the entries that have been allocated).

Time complexity of the operations:

- *Allocate* – zero clock cycles (asynchronous response).
- *Release* – one clock cycle.
- *Check* – zero clock cycles (asynchronous response).

6.2.2 Thread Table

The thread table consists of a data memory only. Its block diagram is shown in Fig. 6.6.

Table 6.1 Family table – field descriptions

RTL name	R/W//R-W	Size (generic name)	UTLEON3 value
tail	1/0/1	TT_A_WIDTH	8
p_fid	1/0/1	FTT_A_WIDTH	5
p_tid		TT_A_WIDTH	8
index.start		FTT_TINDEX_WIDTH	16
index.limit		FTT_TINDEX_WIDTH	16
index.step		FTT_TINDEX_WIDTH	16
index.next_index		FTT_TINDEX_WIDTH	16
reginfo.addr_globals		REG_A_WIDTH	10
reginfo.addr_shared		REG_A_WIDTH	10
reginfo.addr_base		REG_A_WIDTH	10
reginfo.addr_latest		REG_A_WIDTH	10
reginfo.addr_top		REG_A_WIDTH	10
reginfo.num_globals		REGS_PER_THREAD_WIDTH	5
reginfo.num_shared		REGS_PER_THREAD_WIDTH	5
reginfo.num_locals		REGS_PER_THREAD_WIDTH	5
exit_regaddr		REGS_PER_THREAD_WIDTH	5
flags.created		1	1
flags.init_done		1	1
flags.first_thread		1	1
flags.allocation_done		1	1
block_size		TT_A_WIDTH	8
block_used		TT_A_WIDTH	8
pc		32	32

The Data memory contains static and runtime parameters of threads (see Table 6.2). As in the case of the family table, single fields of the thread table are merged into groups. Nevertheless, the ways these fields are accessed require the use of ten independent groups, with only two of them consisting of more than one field. Furthermore, two groups required three ports instead of two. All but the *state* groups are mapped to a dual-port memory soft macro (DP RAM) that is implemented using one or more dual-port BRAMs. Multiple read ports are served by multiple instances of the same memory. Each *state* group is mapped to a *state mem* component that allows to set/clear single bits of the group independently.

6.3   Thread Scheduler

The thread scheduler is responsible for thread management. It is a brand new component in the UTLEON3 processor that does not have a counterpart in the original LEON3 processor.

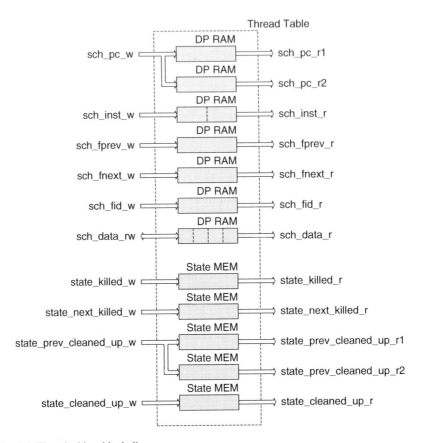

Fig. 6.6 Thread table – block diagram

As shown in Fig. 6.7, the thread scheduler consists of four main functional
blocks:

- *Create* – family initialization and finalization, thread creation
- *Pick* – provides threads to processor pipeline inputs
- *Push* – processes threads from processor pipeline outputs
- *Cleanup* – cleans up finished threads

The functions of the scheduler are:

- Resource allocation (family table entries, thread table entries, registers in the
  register file).
- Update of family parameters and family states.
- Family initialization prior to thread creation.
- Thread creation, using either new thread table entries, or reusing the entry of a
  preceding thread that has already finished.

Table 6.2 Thread table – field descriptions

RTL name	R/W/ /R-W	Size (generic name)	UTLEON3 value
pc	2/1/0	32	32
inst.inst	1/1/0	32	32
inst.flow		2	2
fprev	1/1/0	TT_A_WIDTH	8
fnext	1/1/0	TT_A_WIDTH	8
fid	1/1/0	FTT_A_WIDTH	5
data.reg_base	0/1/1	REG_A_WIDTH	10
data.reg_producer		REG_A_WIDTH	10
data.is_tid_0		1	1
data.is_last_in_family		1	1
data.index		FTT_TINDEX_WIDTH	16
state.killed	1/1/0	1	1
state.next_killed	1/1/0	1	1
state.prev_cleaned_up	2/1/0	1	1
state.cleaned_up	1/1/0	1	1

- Preparation of threads for execution in the processor pipeline.
- Processing of threads at the processor pipeline writeback stage.
- Performing thread wake-ups – both threads that are *waiting* due to an I-Cache miss and threads that are *suspended* due to reading *pending* source registers.
- Cleanup of finished threads for further reuse (thread table entries, registers, TLS).
- Family finalization.
- Resource release.

A block diagram of the thread scheduler and its interfaces to other components is shown in Fig. 6.7. The list of the interfaced components including the functional connections is as follows:

- Family table

    - Update family global parameters and family states (see Table 6.1).
    - Serve requests for family table entries – allocation, release, check.

- Thread table

    - Update thread static and runtime parameters (see Table 6.2).

- I-Cache

    - Request the first instruction of a thread. This response is asynchronous to the request issued by the scheduler.
    - Request the register mask information (.registers) of a family prior to its initialization.

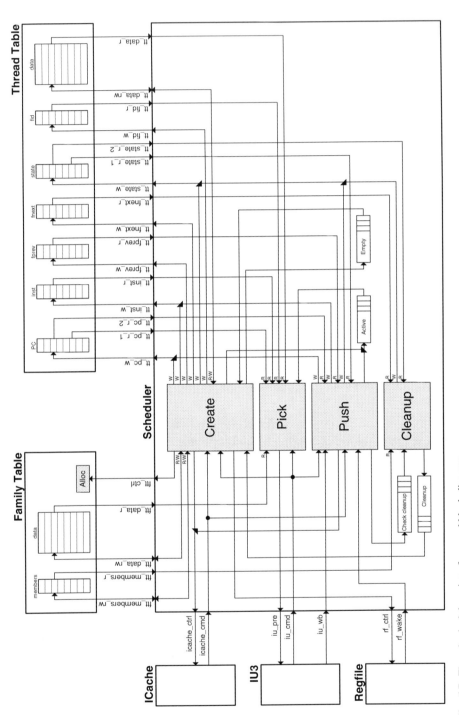

Fig. 6.7 Thread scheduler – interfaces and block diagram

- Integer pipeline (IU3)

  - Provide all the necessary parameters of a thread that is to enter the pipeline to pipeline inputs.
  - Process commands issued from the processor pipeline.
  - Process threads from the processor pipeline *XC* stage.

- Register file

  - Generate requests for allocation of a block of registers.
  - Update register values and states.

The main function of the thread scheduler is thread management. Each thread is identified by an index to a thread table entry (Thread ID, TID) that was assigned to this thread. Thus, the terms thread, TID and thread table entry are used interchangeably in the text below when it does not change the meaning.

Each thread can be set to one of the seven states that are defined in the list below. State transitions and events that caused these transitions are shown in Fig. 6.8.

- *Empty*

  - The thread table entry is not used by any thread.
  - This is an initial value of each thread table entry. Second, a TID becomes *empty* again when it is no longer needed by the family it was assigned to.
  - *Empty* TIDs are stored in the *empty queue* in the scheduler

- *Active*

  - The thread is ready to be executed in the processor pipeline.
  - A thread changes its state from *waiting* to *active* when an *I-Cache wakeup* request occurs. It means that the required cacheline has been loaded in the I-Cache, and the first instruction that the thread should start with has been stored in the *inst* field of the thread table. Second, a *running* thread becomes *active* when an instruction with the *SWCH* modifier was decoded in the processor pipeline. This transition happens at the time the corresponding instruction passes the *XC* stage of the processor pipeline.
  - Active TIDs are stored in the *active queue* in the scheduler.

- *Running*

  - The thread is in the processor pipeline.
  - A thread becomes *running* when it enters the processor pipeline. It remains in this state until an I-Cache miss occurs in the *FE* stage, or the *END* or *SWCH* modifier is decoded in the *DE* stage, or a *PENDING* register is read in the *EX* stage of the processor pipeline. All these transitions happen at the time the corresponding instruction passes the *XC* stage of the pipeline.
  - *Running* threads are not listed explicitly in any queue. They are qualified as *running* by their presence in the processor pipeline instead.

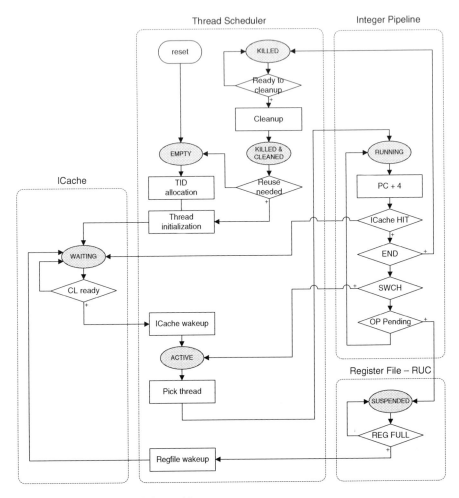

Fig. 6.8 Thread states and their transitions

- *Suspended*
  - The thread is waiting for a register to be updated to the *FULL* state.
  - A thread becomes *suspended* when a *PENDING* register is read in the *EX* stage of the processor pipeline. The source register becomes *WAITING*, and the TID is stored in its data part.
  - TIDs of *suspended* threads are stored in the data part of the register that each thread is waiting for. Note that only one thread can be waiting for a given register.

- *Waiting*
  - The thread is waiting for an instruction (e.g. after an I-Cache miss)

Table 6.3 Thread scheduler – timing of operations

Operation	Throughput [CC per TID]	Latency [CC]	Comment
IU3 interface			
Pick	1 buffered	3	Pipeline + buffer
	2 direct		
Push	1	1	Buffer
Internal operations			
Create (one thread)	1 + 1 + #%ts	1 + 1 + #%ts	Not pipelined
Wakeup from I-Cache	1	1	Buffer
Wakeup from regfile	2	3–4	Buffer + pipeline
Cleanup	2	1 first thread	Not pipelined
		2 subsequent threads	

- A thread becomes *waiting* when an instruction it should start with is requested from the I-Cache (thread initialization, register file wake-up), or if an I-Cache miss occurs in the FE stage of the processor pipeline.
- TIDs of *waiting* threads are stored in internal data structures in the I-Cache

- *Killed*

  - The thread has finished
  - A thread becomes *killed* when the *END* modifier is decoded in the DE stage of the processor pipeline.
  - Finished threads are identified by the *state.killed* flag in the thread table.

- *Killed & cleaned up*

  - The thread has finished, and the thread table entry is ready to be reused by another thread from the same family or released for another family
  - A thread becomes *killed & cleaned up* when the previous thread is *killed & cleaned up* and the next thread is *killed*.
  - *Killed & cleaned up* TIDs are stored in the *cleanup queue* in the scheduler.

The timing of the basic scheduler operations is listed in Table 6.3. To explain the meaning of the terms in the table:

Pick – the thread scheduler holds two threads in a buffer in the *Pick* stage to be able to serve two subsequent thread switch requests with zero latency. In case of more requests in a line the *Pick* stage provides a new thread every other clock cycle. Once the requests end, the *Pick* stage restores the buffer, and then again it is able to serve thread switch requests with zero latency.

Push – the *Push* stage is able to process threads in the rate of one thread per clock cycle.

Pipeline – in the table means the interface (or operation) is pipelined. The processing of the first item takes the *"Latency"* clock cycles, but subsequent items are produced in a rate shown in the *"Throughput"* column.

Buffer – denotes that a buffer of a sufficient length is used at the interface to avoid blocking the producer when inserting wait cycles on the consumer side of the interface.

Not pipelined – the interface or operation is not pipelined.

Pipeline + Buffer – a pipeline and buffer can be combined in one interface. The *Pipeline + Buffer* combination means that the data are passed through the pipelined processing first and are then stored into the buffer to be picked by the consumer.

Buffer + Pipeline – the Buffer + Pipeline combination means that the data are put into the buffer by the producer and then picked from the buffer and processed in the pipeline.

6.3.1 Processor Pipeline

The original LEON3 processor pipeline was extended to be able to process microthreaded code. The pipeline can work in two modes – the legacy mode and the microthreaded mode. When in the legacy mode, the pipeline is backward compatible with the original LEON3 processor pipeline. In the microthreaded mode all the microthreaded extensions are enabled, and the pipeline is able to process the microthreaded code. The pipeline is switched to the microthreaded mode when the launch instruction is decoded, and back to the legacy mode when *thread 0* ends.

The microthreaded extensions of the processor pipeline are shown in Fig. 6.9. Unlike the original processor and the legacy mode of the UTLEON3 processor, where a single stream of instructions is processed, multiple threads are active concurrently in the microthreaded mode, and their instruction streams are interlaced in the processor pipeline. The thread scheduler is responsible for providing threads to the pipeline inputs and capturing threads at the pipeline outputs in order to update their state. This is performed through three interfaces:

- sch_pre – the input for threads that are to be executed in the pipeline. It consists of

  - Thread identifier (TID)
  - Program counter and prefetched instruction that the thread should start with
  - Address and numbers of registers that the thread can access
  - Additional parameters and flags
  - Nullification request *force_nop*, asserted when no threads are *active*

- sch_wb – the output for threads that leave the pipeline. It consists of

  - Thread identifier (TID)
  - Program counter that the thread should start with the next time it enters the pipeline
  - Thread state (*EMPTY, RUNNING, SWITCHED, SUSPENDED, WAITING, KILLED*)

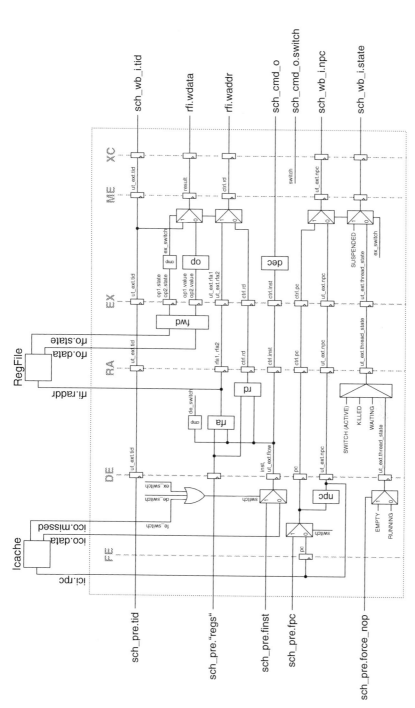

Fig. 6.9 Microthreaded extensions of the processor pipeline

- sch_cmd – commands from the pipeline to the scheduler

    – Commands derived from the decoded instructions (launch, allocate, set..., create)
    – Request for a thread switch
    – Additional parameters

The three processor pipeline stages most influenced by the microthreaded extensions are

- Fetch stage (FE)
- Decode state (DE)
- Execute stage (EX)

The *fetch stage* computes an address of the next instruction, and sets the initial
state of a thread.

The fetch stage computes an address of the next instruction, provides this address
to I-Cache inputs, and passes this address (i.e. the program counter) along with
the instruction returned by the I-Cache to the decode stage one clock-cycle later.
Unlike the legacy mode, in which the processor pipeline is stalled on an I-Cache
miss, the thread switch (*FE_SWITCH*) is asserted in the microthreaded mode on
an I-Cache miss, and a program counter value and an instruction of the next active
thread are provided to the decode stage instead. This allows to perform thread switch
on an I-Cache miss with a zero clock-cycle latency (i.e. no bubbles are inserted in
the pipeline). Note that the next program counter (*npc*) is provided to the decode
stage as well. The program counter (*pc*) refers to the current instruction (*inst*) which
will be the last instruction that the thread will process when a thread switch occurs.
The next program counter (*npc*) refers to the instruction that the thread will continue
with the next time it enters the pipeline.

The thread state is set as *RUNNING* by default, provided that an active thread
is ready at the pipeline input. The thread state is set to *EMPTY* otherwise, which
means that the slot in the pipeline will be nullified. Nevertheless the pipeline is
never stalled.

The *decode stage* decodes the 2-bit instruction modifier, computes register
addresses, and updates the thread state.

The 2-bit modifier is decoded, and if the value *SWCH* or *END* is recognised, a thread switch (*DE_SWITCH*) is asserted. As in the case of *FE_SWITCH*,
DE_SWITCH is performed with a zero clock-cycle latency.

Register addresses are computed from the parameters (base addresses and
numbers of registers) provided by the scheduler and the RS1, RS2 and RD fields
of the instruction. Unlike the legacy mode, the window pointer *cwp* is not used.

The thread state is set to *SWITCHED* (*KILLED*) when a *SWCH* or (*END*)
instruction modifier is recognized. The thread state is set to *WAITING* when an
I-Cache miss occurs for the following instruction of the thread. It means that the
instruction in the decode stage will be the last instruction of the thread, and the
thread can continue after the cacheline that the following instruction belongs to has

been loaded in the I-Cache. Furthermore, this instruction will be stored in the thread scheduler so that it can provide it at the pipeline input at the time the thread enters the pipeline again.

The *execute stage* checks states of the instruction operands.

If at least one of the operands is *PENDING*, a thread switch (*EX_SWITCH*) is asserted. In this case the decode (DE), register access (RA) and execute (EX) stages have to be annuled, which introduces a three-cycle long bubble in the pipeline. The register state is changed from *PENDING* to *WAITING*, and the thread index (TID) is written into this register. The current program counter value (pc) is stored in the next program counter value field (npc), since this instruction has to be reissued at the time the *WAITING* source register becomes *FULL*.

The thread state is set as *SUSPENDED* when an (*EX_SWITCH*) occurs.

6.4 Data Cache and Register Update

The data memory access mechanism required by the microthreaded model is implemented in two parts. The first part is a new data cache controller that reads requests from the integer pipeline, and implements memory operations to guarantee data availability and accuracy. The second part is a register update controller (RUC) that reads specific data cache line locations and updates the corresponding *PENDING* or *WAITING* registers in the integer register file. The operation of the update controller is controlled and triggered by the data cache controller.

6.4.1 Data Cache Subsystem

The data cache subsystem consists of four control units (see Fig. 6.10):

- *D-Cache controller* that interacts with the integer pipeline and issues cache-line requests and pending store requests to the D-Cache memory controller.

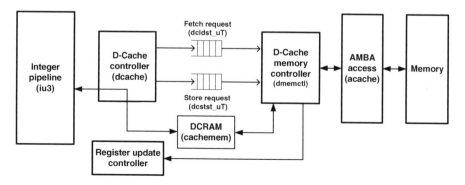

Fig. 6.10 D-Cache subsystem – block diagram

- *D-Cache memory controller* reads pending cache line fetch requests and pending store requests, and fetches the data from the memory and modifies it accordingly.
- *D-Cache line fetch request FIFO* stores pending fetch requests.
- *D-Cache pending store request FIFO* stores pending store operations to be executed on cache line fetch completion by the memory controller.

The cache RAM memory is organized as follows. The data part is implemented as one dual-port memory block. The tag part is regrouped and decomposed into several dual-port memory blocks; this is because both the data cache controller and data cache memory controller access different tags (at different offsets) at the same time (see Fig. 6.11).

Parts accessed by the data cache controller:

- *ST* (R/W) – the current state of a cacheline as seen by the data cache memory controller set to *valid* on completion of register updates from the cache line.
- *TAG* (R) – the last valid tag (higher address bits) to support writeback.
- *STREQ* (R/W) – the current state of a cacheline as seen by the data cache controller – *valid* or *loading* (requested).
- *TAGREQ* (R/W) – the higher address bits as seen by the data cache controller, updated on a cache miss.
- *HEAD* (W) – the head of the linked list of registers that request a given cache line (RUC init).
- *TAIL* (R/W) – the tail of the linked list of registers that request a given cache line (RUC init, append).
- *WRITEMASK* (R/W) – flags that mark dirty cache words (store first, writeback).

Parts accessed by the data cache memory controller:

- *ST* (R/W) – the current state of a cacheline as seen by the data cache memory controller.
- *TAG* (R/W) – the last valid tag (higher address bits) to support writeback.
- *STREQ* (R/W) – the current state of a cacheline as seen by the data cache controller.
- *TAGREQ* (R/W) – the requested tag (i.e. new tag of the cache line to be fetched).
- *HEAD* (R) – the head of the linked list of registers that request a given cache line (RUC update).
- *TAIL* (R) – the tail of the linked list of registers that request a given cache line (RUC update).
- *WRITEMASK* (W) – flags indicating the dirty cache words (writeback).

The state transition diagram for data cache lines is shown in Fig. 6.12. After power-on all cache lines are marked as *invalid*. On a cache line fetch request the corresponding cache line is marked as *loading*. On a fetch completion the line is either marked as *valid* when no register updates have been registered for this cache line and the corresponding store queue is empty, otherwise it is marked as *updating*

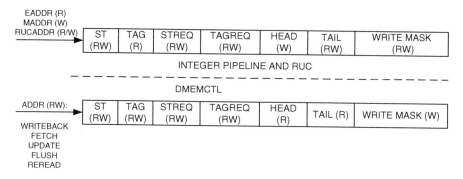

Fig. 6.11 Data cache – tag access

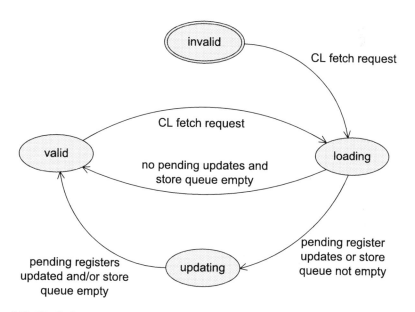

Fig. 6.12 The D-Cache line status transitions

until the register updates and/or store queue have been processed. The store queue is implemented as a set of 16 independent FIFOs (dcstst) where each FIFO maps to exactly one pending cache line fetch request stored in another FIFO (dcldst).

A simplified diagram of the data cache controller is shown in Fig. 6.13, and the data cache memory controller is shown in Fig. 6.14. Both parts do not synchronize their execution directly, they communicate through the load and store FIFOs and through the tag part of the cache line stored in the cachemem block. The RUC update structures are initialized and filled in by the data cache controller, and the actual register update sequence is triggered by the data cache memory controller on cache line fetch completion.

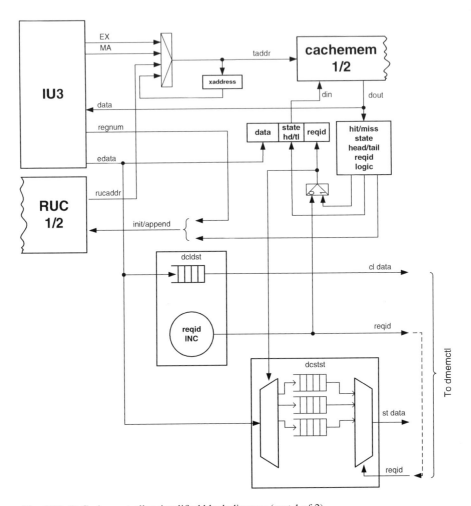

Fig. 6.13 D-Cache controller simplified block diagram (part 1 of 2)

As the data cache on the pipeline side provides shared access to the integer pipeline and the register update controller, it is necessary to guarantee proper address generation when switching between various modes of operation. In the normal operating mode the access is granted to the integer pipeline before the register update controller. When the pipeline is recovering from a stall (*holdn* active), it is necessary to guarantee completion of possible write operations performed at the end of each RUC update cycle (change of line status). In these situations one or more additional wait states must be inserted; this is caused by sharing a single port of a dual-port BRAM memory for this access (the other is used on the memory side of the D-Cache controller) and by the one clock cycle latency of the memory. The most

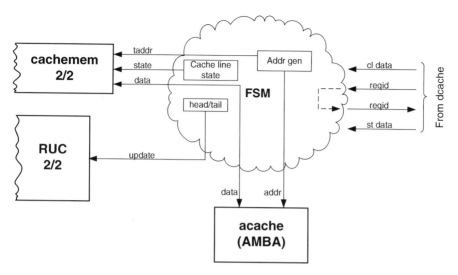

Fig. 6.14 D-Cache controller simplified block diagram (part 2 of 2)

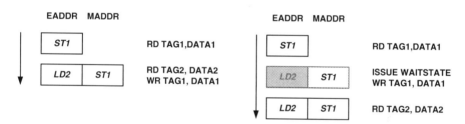

Fig. 6.15 D-Cache example 1 – normal pipeline access

important situations are shown in Figs. 6.15–6.17. In each figure the left part shows an ideal situations where we can process each memory access instruction in one clock cycle, and the right part shows the real implementation in UTLEON3.

The first situation – an *ST* instruction followed by an *LD* instruction – requires an additional wait state to be inserted (pipeline stalled for one clock cycle) to finish the read/write cycle of the first instruction before the next address can be applied to the cache BRAM.

The second situation extends this further by halting the pipeline after the first clock cycle of the *ST* instruction. When *holdn* is active, RUC can access the cache BRAM at any time, thus it is necessary to guarantee a proper pipeline restart when *holdn* goes inactive by inserting one additional wait state as in the previous case.

This is further complicated in the third situation where in the last cycle of *holdn* active RUC finishes an update round, which requires the corresponding tag be updated. In this case three additional wait states have to be inserted, one to modify the tag due to RUC, one to restart the *ST* operation and one to write the new data in the *ST* operation.

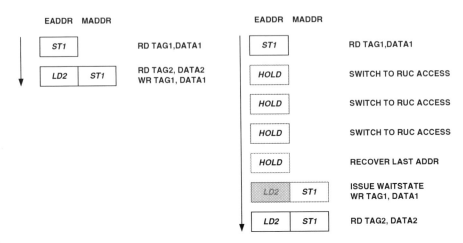

Fig. 6.16 D-Cache example 2 – pipeline access with an interleaved RUC access due to *holdn* active

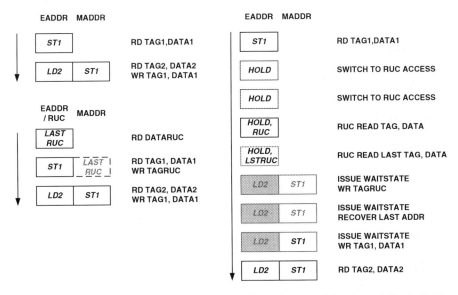

Fig. 6.17 D-Cache example 3 – pipeline access with an interleaved RUC access due to *holdn* active with RUC done active in the last active cycle of *holdn*

From the examples it can be seen the data cache controller on the pipeline side operates in these modes:

• The address is taken from the pipeline execute stage.
• The address is taken from the pipeline memory access stage.

- The address is taken from the pipeline memory access stage and a wait state is generated (the pipeline is stalled for one clock cycle). This happens only on a cache miss, on accessing different cache lines, when ST is immediately followed by LD/ST, or on LDD/STD
- The address is taken from the RUC interface when *holdn* is active or the pipeline does not access the cache RAM.
- The address is taken from the RUC interface when *holdn* is inactive and this is the last access in a register update batch.

6.4.2 Register Update Controller

The register update controller (RUC) consists of four blocks. Its structure is shown in Fig. 6.18:

- *Decoder* – is a combinational logic that receives *init* and *append* requests from the data cache controller, and translates them to operations on the linked list.
- *Linked list* – is a data structure based on two dual-port memory blocks that store *(register ID, cache line ID)* pairs. The *register ID* is a pointer to the next entry, and the *cache line ID* identifies the location of the data within the cache memory. Two memory blocks are required as two different list items (two memory addresses) are accessed at the same time.
- *Update request block* – is simply a FIFO that stores update requests formed by *(head, tail)* pairs issued by the data cache memory controller.
- *RUC FSM* – is a finite state machine that reads the cached data and executes register writes based on the linked list.

The RUC executes three operations: *init, append* and *update. Init* and *append* modify the linked lists (one list per cache line), *update* triggers a regfile update. The operations are further explained below:

- *Init* – The *tail* item addressed by the register ID is set with the current register ID. The cache line ID item is set with the current values (see Fig. 6.19).

```
head = regnum
tail = regnum
BRAM1[regnum] = tail (i.e. Register ID)
BRAM2[regnum] = (set, offset, line) (i.e. Cache
                       line ID)
```

- *Append* – The *tail* item addressed by the register ID is not set with any value. The *tail* item addressed by the previous register ID is set with current register ID. The cache line ID item is set with current values (see Fig. 6.20).

```
tail = regnum
BRAM1[prev_regnum] = tail
BRAM2[regnum] = (set, offset, line)
```

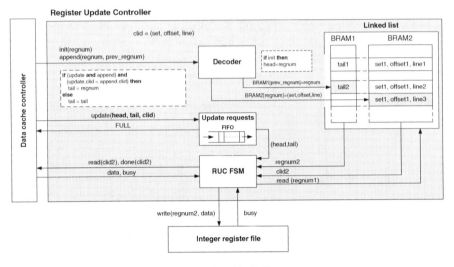

Fig. 6.18 The register update controller (RUC) – block diagram

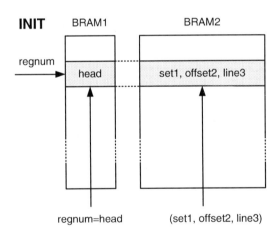

Fig. 6.19 The register update
controller (RUC) – operation
init

- *Update* – The *head* and the latest *tail* are stored in the update request structure. Both values are register IDs. The RUC FSM generates requests given by the relevant linked list entries to the data cache controller and receives the valid data. The data are written to the integer register file (see Fig. 6.21).

6.5 Instruction Cache

The instruction cache receives requests for memory access from three places: the integer pipeline fetch stage as part of the instruction fetch, the integer pipeline

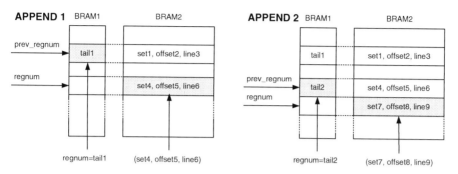

Fig. 6.20 The register update controller (RUC) – operation *append*

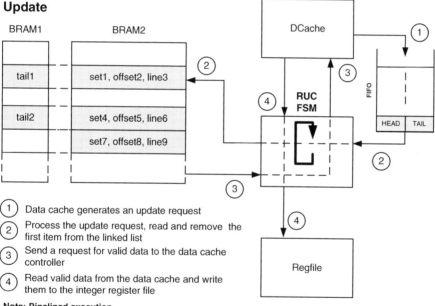

Update

1 Data cache generates an update request
2 Process the update request, read and remove the first item from the linked list
3 Send a request for valid data to the data cache controller
4 Read valid data from the data cache and write them to the integer register file

Note: Pipelined execution

Fig. 6.21 The register update controller (RUC) – operation *update*

execute stage when the *setthread* instruction is executed, and the thread scheduler. The structure of the instruction cache subsystem is shown in Fig. 6.22.

The instruction cache on the pipeline side must support the following concurrent operations:

1. Fetch instructions – generate hit/miss, pass the flow information (2-bit instruction modifiers) to the scheduler, generate CL fetch requests on misses.
2. Execute *setthread* – pass .registers to the scheduler, generate CL fetch requests on misses.

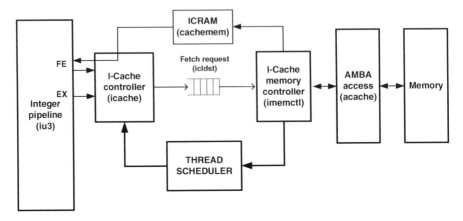

Fig. 6.22 I-Cache subsystem – block diagram

3. Increment and decrement reference counters as threads enter and leave I-Cache lines.
4. Annul IU3 CL fetch requests in the EX stage that were generated in the FE stage for the instruction in the EX stage in situations where unsatisfied data dependencies in the thread have been detected prior to the instruction that generated the cache miss.
5. Test instruction availability for the scheduler – generate a hit/miss response, generate a CL fetch request on miss. This is used at the beginning of thread creation to guarantee all threads have their I-Cache lines prefetched. In the current implementation this request always results in an I-Cache miss; this is to relieve the pressure on the pipeline side BRAM port by moving these requests to the memory side BRAM port.

On the memory side the I-Cache controller must perform the following exclusive operations:

• Generate a thread wake up request on I-Cache line fetch completion. This is a response to an I-Cache miss generated on instruction fetch (step 1 in the previous list).
• Pass .registers to the scheduler. This is a response to executing *setthread* (step 2 in the previous list).
• Pass an instruction and flow (2-bit instruction modifier) to the scheduler. This is a response to the operation in step 5 in the previous list.

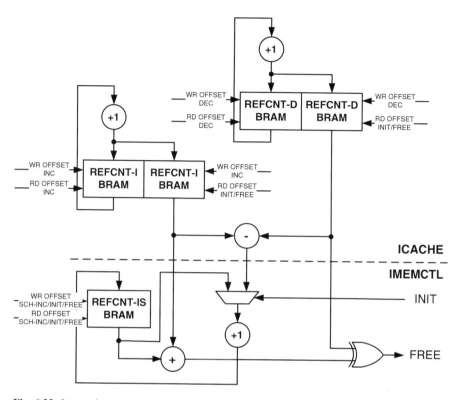

Fig. 6.23 Instruction cache – reference counter implementation

6.5.1 Reference Counters

In the microthreaded model the coherence of thread scheduling and instruction cache is maintained via *reference counters* that store for each cache line the number of threads that are currently using it. Each reference counter is incremented when a thread enters a cacheline (both on cache hit and miss). Each reference counter is decremented when a thread leaves a cacheline (always hit – by definition).

The instruction cache subsystem must be able to increment and decrement at least two independent thread counters in one clock cycle (instruction fetch). Furthermore, thread counters can be incremented on execution of the *setthread* instruction, and by a scheduler query when marking threads ready for execution. Given the limitations of the current FPGA technology (dual-port block RAM primitives), this leads to the architecture shown in Fig. 6.23.

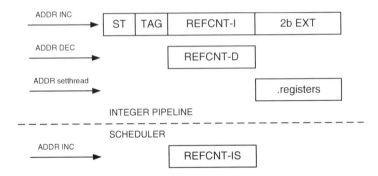

Fig. 6.24 Instruction cache – tag access

It can be seen that a reference counter for one cache line is computed out of three independent reference counters – incrementors:

- *REFCNT-I* – is the number of reference counter increments generated by the integer pipeline (both in the fetch and execute stages).
- *REFCNT-D* – is the number of reference counter decrements generated by the integer pipeline (in the fetch stage).
- *REFCNT-IS* – is the reference counter offset written by the instruction cache memory controller on completion of a cache line fetch. Its initial value is calculated as a sum of *REFCNT-I* and *REFCNT-D* counters at the time of the cache line fetch incremented by one; this reduces the need to reset the two other reference counters, and hence reduces the number of necessary write ports. This value is also incremented on scheduler requests.

6.5.2 Tag Access

Figure 6.24 shows tag organization and possibilities of independent access. At one time the tag can be requested from the instruction cache controller side independently by the

1. Integer pipeline *fetch stage*,
2. Integer pipeline *execute stage*,
3. *Thread scheduler*.

As the implementation allows only one access from the instruction cache controller side, these requests are arbitrated in the order shown above.

The tag contains two other fields that store the cache line word #0 that corresponds to the 2-bit extensions (instruction modifiers) of the microthreaded instruction set, and to word #1 that corresponds to the possible `.registers` value required by the scheduler when launching new threads. Word #0 is accessed

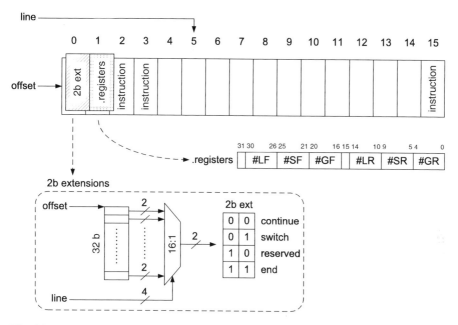

Fig. 6.25 Instruction cache – cache line storage organization

instantly, and it is desirable to decrease the latency when accessing word #1. These fields are duplicated in the instruction cache line to maintain compatibility with the legacy LEON3 code. The instruction cache memory organization is shown in Fig. 6.25.

Chapter 7
Execution Efficiency of the Microthreaded Pipeline

When analyzing execution efficiency of the microthreaded pipeline, we are interested in two key things:

1. The number of stall clock cycles in the processing pipeline.
2. The latency tolerance for blocking and non-blocking long-latency operations.

This chapter will discuss these topics and present UTLEON3 performance data for simple legacy and microthreaded assembler programs executed in UTLEON3.

7.1 Pipeline Execution Profile

We begin by analysing executions of simple DSP kernels on the original LEON3 processor and on the UTLEON3 processor, both in the legacy mode (L3) and in the new microthreaded mode (UT).

Sources of delays (latencies) in the legacy program execution are described in the first part. The analysis is valid both for the original LEON3 processor and the UTLEON3 processor in the legacy mode. A simple example of a dot-product kernel is used for the demonstration. Afterwards the example is implemented in the microthreaded (UT) mode to show how microthreading can overcome the delays.

In the second part pipeline execution profiles are evaluated experimentally using three DSP kernels. A comparison of an overall program runtime and a breakdown of the pipeline execution profile is provided as well.

It will be shown that the execution of a microthreaded code on the UTLEON3 processor is 1.86 faster on average than the execution of the legacy code on the original LEON3 processor, and 1.66 times faster than the execution of the legacy code on the UTLEON3 processor. Furthermore, a single UTLEON3 core achieves 0.83 instructions per cycle (IPC) on average in the microthreaded mode, but only 0.52 IPC on average in the legacy mode. This indicates a higher efficiency of the microthreaded mode.

M. Daněk et al., *UTLEON3: Exploring Fine-Grain Multi-Threading in FPGAs*,
DOI 10.1007/978-1-4614-2410-9_7, © Springer Science+Business Media, LLC 2013

7.1.1 Sources of Delays in Program Execution

Program runtime in terms of an overall number of clock cycles needed for program execution depends on the number of instructions that the program is composed of and the number of additional delays imposed by data dependencies. In the legacy mode, the processor pipeline is stalled when an unresolved data dependency occurs. Microthreading should overcome this issue as the code is split into simple threads that can execute independently. If an unresolved data dependency occurs, a thread switch is performed, and another thread is executed instead. Data dependencies can be either external, defined by external components such as the memory subsystem, or internal, defined by the internal architecture of the processor pipeline.

External data dependencies are one source of pipeline stalls. A typical cause is the latency of the memory subsystem. If the ld instruction is being processed in the legacy mode, the pipeline is stalled until the cache subsystem returns the required data. The microthreaded mode deals with this issue through a thread switch that is performed whenever a pipeline stall can occur. More precisely, the thread switch is not performed at the time the ld instruction is executed, but when the data required by a subsequent instruction are not available. The switch happens automatically, the thread affected by a cache miss is suspended until the corresponding data are loaded, and the processor pipeline executes other threads in the meantime.

The latency due to external data dependencies cannot be evaluated by a static analysis of the program code. Nevertheless, its impact on the overall execution time can be significant, especially when processing sparse data arrays which may induce high cache miss rate.

Internal data dependencies are another source of pipeline stalls. They are defined by the instruction set architecture (ISA) and structure of the data paths in the processor.

Some constraints imposed by the LEON3 processor architecture are:

- The distance between the ld instruction and the subsequent instruction that depends on its result must be three slots at least. The bubbles (i.e. unused slots) in processor pipeline are induced otherwise, resulting in up to 2CC long delay.
- The distance between the MUL instruction and the subsequent instruction that depends on its result must be two slots at least. As in the previous case, shorter distances result in a bubble 1CC long.
- The distance between an arithmetic instruction that sets the integer condition codes (ICC) and a branch instruction must be at least three slots. As in the previous case, shorter distances result in a bubble up to 2CCs long.
- A MUL instruction always generates a bubble 4CCs long.

The first three constraints mentioned above are imposed by internal data dependencies between two instructions. They can be eliminated by rearranging the instructions in the source code, but this technique may not be always applicable as will be shown later.

```
      ld [%r1 + %r8], %r10    /* %r10 = a[i] */
   2: ld [%r2 + %r8], %r11    /* %r11 = b[i] */
      subcc %r8, 4, %r8       /* i -= 4 */
      umul %r10, %r11, %r10   /* +(1+4)CC bubble /*
      add %r10, %r9, %r9      /* +1CC bubble /*
      bpos,a 2b               /* if (i > 0) goto 2: */
      ld [%r1 + %r8], %r10    /* %r10 = a[i] (delay slot) */
```

Fig. 7.1 Legacy assembler code of an optimized dot-product loop. The comments reflect the bubbles generated by particular instructions

The microthreaded mode also cannot deal with these internal data dependencies implicitly, but they can be eliminated through an explicit thread switch by appending the swch instruction modifier to the producer instruction of the data-dependent instruction pair.

The last constraint in the list above is caused by the latency of the integer multiplier. In the legacy mode the processor pipeline has to wait until the multiplier provides the result that can be written to the destination register.

On the contrary, in the microthreaded mode multiplication is treated as a long-latency operation. This means that the destination register is set to pending when the MUL instruction is issued, and it is updated later when the result is available. This approach supports multiplier pipelining implicitly, which increases the utilization of the integer multiplier.

We can conclude that all the constraints mentioned above can be fully eliminated in the microthreaded mode, which leads to reduction of program execution time and higher instruction-per-cycle values. The following case study analyzes a simple dot-product loop.

7.1.1.1 Program Execution in the Legacy Mode

Figure 7.1 shows a dot-product in the assembler code that has already been optimized in order to reduce the number of bubbles in the processor pipeline. Despite the optimization, one iteration of the loop still generates 6CC bubbles: 2CCs are due to inter-instructions dependencies; 4CCs are due to the multiplier latency. As the length of the loop body is equal to six instructions only, these bubbles represent 50 % of the overall loop-iteration runtime (i.e. 12CC) provided that only internal data dependencies are taken into account.

The problem of data dependencies can be partially reduced by loop unrolling. A higher number of instructions within a loop body allows to arrange the instructions in a more efficient way that leads to a more efficient code. Figure 7.2 shows a dot-product loop with an unroll factor 2. The bubbles caused by the inter-instructions dependencies were eliminated completely. The remaining 8CC bubbles are caused by the latency of the multiplier. As the length of the loop body is equal to 12 instructions, the proportion of bubbles in the overall loop-iteration runtime (i.e. 20CC) was reduced to 40 %.

```
       add %r1, %r8, %r20      /* a+i */
2:  add %r2, %r8, %r21      /* b+i */
       ld [%r20 + 0], %r10     /* %r10 = *((a+i)+0) */
       ld [%r20 + 4], %r12     /* %r12 = *((a+i)+4) */
       ld [%r21 + 0], %r11     /* %r11 = *((b+i)+0) */
       ld [%r21 + 4], %r13     /* %r13 = *((b+i)+4) */
       subcc %r8, 2*4, %r8     /* i -= 8 */
       umul %r10, %r11, %r10 /* +4CC bubble /*
       umul %r12, %r13, %r12 /* +4CC bubble /*
       add %r10, %r9, %r9
       add %r12, %r9, %r9
       bpos,a 2b               /* if (i > 0) goto 2: */
       add %r1, %r8, %r20      /* a+i (delay slot */
```

Fig. 7.2 Legacy assembler code of a dot-product loop, optimized by loop unrolling with unroll factor 2

```
       set a, %l1              /* set %g1 */
       set b, %l2              /* set %g2 */
       set 0, %l3              /* clear %d0 */

       allocate %l4            /* allocate a family */
       setlimit %l4, %g3       /* set upper bound */
       setstep %l4, 4          /* set cycle step */
       set dp_uthread, %l5     /* get address of the code */
       setthread %l4, %l5      /* set start address */
       create %l4, %l4         /* request family creation */

       mov %l4, %l5 ; SWCH     /* sync */
```

Fig. 7.3 Microthreaded assembler code for initialization of a family that computes a dot-product

An additional loop unrolling still improves the code runtime since the loop control code of a fixed length becomes shared by more loop iterations. Nevertheless the number of bubbles is still proportional to the number of the MUL instructions.

7.1.1.2 Program Execution in the Microthreaded Mode

Microthreading can eliminate the issues mentioned in the previous text completely. The original loop is replaced by a so-called family of threads; each iteration of the loop is represented by a single thread. Thread management that replaced the loop control completely is performed by the hardware thread scheduler. A microthreaded assembler code for initialization of a family of threads is shown in Fig. 7.3; A microthreaded assembler code for the dot-product computation is shown in Fig. 7.4.

The number of instructions that are needed to compute one result value is reduced since the loop control instructions are not present in this code. The thread index is provided in the first local register of the thread; this register is set by the hardware thread scheduler when a thread is in the initialization phase.

```
.registers 3 1 3      /* GR, SR, LR */
/* %g1 = a */
/* %g2 = b */
/* %10 = i */
/* %d0 = sum(i-1) */
/* %s0 = sum(i) */
dp_uthread:
  ld [%g1+%10], %11              /* %11 = *(a+i) */
  ld [%g2+%10], %12  ; SWCH  /* %12 = *(b+i) */
  umul %11, %12, %12  ; SWCH
  add %d0, %12, %s0   ; END   /* sum += %12 */
```

Fig. 7.4 Microthreaded assembler code for a thread that performs a dot-product computation

swch modifiers are used to eliminate stalls imposed by data dependencies. The swch behind the second ld instruction prevents a 2CC bubble due to the inter-instructions data dependency (the ld-umul instruction pair). Furthermore, it gives more time to the cache-subsystem to update the destination register of an ld instruction on a D-Cache miss before it is requested by the next instruction. Nevertheless, if the thread re-enters the pipeline and the data are still not available for the umul instruction, a regular thread switch occurs and the thread is suspended until the data are ready. This is the only case that results in a 1CC bubble; no bubbles are induced otherwise. A swch behind the umul instruction forces another explicit thread switch. As the umul instruction is a long-latency operation, the succeeding add instruction would cause a regular switch anyway. By the use of the explicit switch proactively, the thread is switched and is likely to re-enter the pipeline after the destination register of umul instruction had been updated, thus the add instruction can be executed immediately. No bubbles are induced as a result.

As the thread code (Fig. 7.4) consists of four instructions and there are no pipeline bubbles, except the single case mentioned above, the static runtime of a single thread, which corresponds to one loop iteration, is just 33 % of the runtime required by the original legacy code.

The loop initialization consists of four instructions in the case of the legacy code, and ten instructions (see Fig. 7.3) in the case of the microthreaded code. Its impact on the overall loop runtime depends on the number of loop iterations, but it can be considered as negligible.

7.1.2   Experimental Results

This subsection provides results of two experiments. The first one deals with three various DSP kernels, the second one deals with the impact of an unroll factor of the FIR filter inner loop to program runtime.

Table 7.1 Pipeline execution profile for three DSP kernels, executed in the legacy mode (L3) and the microthreaded mode (UT)

Core	Program	Total [1]	Working [1]	[%]	Holdn(cache) [1]	[%]	Stall [1]	[%]	UT overhead [1]	[%]
LEON3	L3 FIR 4x	7,477	4,105	55	268	4	3,104	42		
UTLEON3	L3 FIR 4x	7,421	4,105	55	212	3	3,104	42		
UTLEON3	UT FIR 4x	4,210	3,890	92	14	0	98	2	208	5
LEON3	L3 MATRIX	1,717	692	40	337	20	688	40		
UTLEON3	L3 MATRIX	1,586	692	44	228	14	688	43		
UTLEON3	UT MATRIX	900	665	74	0	0	57	6	178	20
LEON3	L3 DCT	6,634	2,920	44	1,854	28	1,860	28		
UTLEON3	L3 DCT	5,066	2,920	58	286	6	1,860	37		
UTLEON3	UT DCT	3,493	2,890	83	22	1	286	8	313	9

Every DSP kernel was hand-coded in assembler in two variants – as the original legacy code (L3) and as the new microthreaded code (UT). These two variants are equivalent in terms of generated data outputs. Three available combinations of the processor core and a program version are considered as follows.

- LEON3-L3 – a legacy program on the original LEON3 processor
- UTLEON3-L3 – a legacy program on the new UTLEON3 processor
- UTLEON3-UT – a microthreaded program on the new UTLEON3 processor

UTLEON3 was configured for these experiments as follows: the family thread table contained 32 rows, the thread table contained 256 rows, the register file contained 1,024 registers and two write ports, and the D-Cache consisted of 2 sets, 1 kB each.

7.1.2.1   Various DSP Kernels

This section shows experimentally evaluated pipeline execution profiles for the following DSP kernels:

- FIR 4x – 26-tap FIR filter with 4x unroll factor
- MATRIX – 4×4 matrix-vector multiplication
- DCT – 2-D 8×8 forward DCT from JPEG

The experimental results for all the three kernels are shown in Table 7.1. The first column shows the processor core; the second column contains the version (L3 vs. UT) and the name of the DSP kernel. The third column contains the overall clock cycle counts. The next eight columns contain a pipeline execution profile breakdown in the form of four couples, each containing the number of clock cycles and the corresponding share in the overall runtime. Four processor pipeline states were considered:

- Working – the processor pipeline is processing useful instruction
- Holdn(cache) – the processor pipeline is idle due to external data dependencies

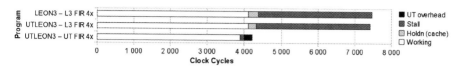

Fig. 7.5 Pipeline execution profile for the 26-tap FIR filter kernel

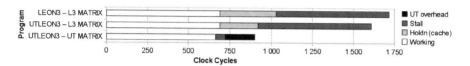

Fig. 7.6 Pipeline execution profile for the 4 × 4 matrix-vector multiplication kernel

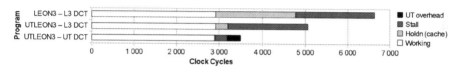

Fig. 7.7 Pipeline execution profile for the 2-D 8 × 8 forward DCT kernel

- Stall – the processor pipeline is idle due to internal data dependencies
- UT overhead – the processor pipeline is idle due to thread management overhead

The processor pipeline state was evaluated as instructions passed through the XC stage of the processor pipeline.

The same numbers are shown graphically in Fig. 7.5 for the 26-tap FIR filter example, Fig. 7.6 for the 4 × 4 matrix-vector multiplication example and Fig. 7.7 for the 2-D 8 × 8 forward DCT.

First, if we compare execution of the same code on both processor cores, UTLEON3 (in the legacy mode) was 1.13 times faster on average. This was due to the different cache-update strategy which reduces the delays caused by external data dependencies (see the column *Holdn*). The numbers in the *Working* column should be equal as the same code was executed. The numbers in the *Stall* column should be equal as the internal data dependencies are handled in the same way.

Second, if we compare execution of both program versions on the UTLEON3 processor, the execution of the UT code was 1.66 faster on average than the execution of the corresponding L3 code. This is due to the elimination of both internal and external data dependencies as was shown on the theoretical example in Sect. 7.1.1.

The further pipeline execution profile breakdown shows the lower number of instructions needed to perform a given task and a significantly smaller overall number of pipeline stalls for the microthreaded code.

The most significant reduction occurs for stalls due to internal data dependencies (see the column *Stalls*). The reason for this was explained on the simple dot-product

Table 7.2 Pipeline execution profile for a 26-tap FIR filter

Core	Program	Total [1]	Working [1]	[%]	Holdn(cache) [1]	[%]	Stall [1]	[%]	UT overhead [1]	[%]
LEON3	L3 FIR 1x	8,976	4,873	54	231	3	3,872	43		
UTLEON3	L3 FIR 1x	9,703	4,873	50	190	2	4,640	48		
UTLEON3	UT FIR 1x	5,877	3,506	60	16	0	107	2	2,248	38
LEON3	L3 FIR 2x	8,230	4,873	59	253	3	3,104	38		
UTLEON3	L3 FIR 2x	8,189	4,873	60	212	3	3,104	38		
UTLEON3	UT FIR 2x	4,612	4,274	93	12	0	104	2	222	5
LEON3	L3 FIR 4x	7,477	4,105	55	268	4	3,104	42		
UTLEON3	L3 FIR 4x	7,421	4,105	55	212	3	3,104	42		
UTLEON3	UT FIR 4x	4,210	3,890	92	14	0	98	2	208	5

example above. It was not eliminated even more mainly due to the presence of two-cycle store instructions, whose first cycle is classified as working, while the second cycle as stalled.

A reduction of latencies due to external data dependencies (see the column *Holdn (cache)*) is not so obvious in this experiment. It is mainly due to the fact that an external memory setup with a short latency of 1CC was used. It is expected that while the runtime of legacy code lengthens with an increasing memory latency, the runtime of the microthreaded code should not change. Experiments with longer memory latencies will be shown later.

The UT versions contains an additional overhead due to thread management (see the column *UT Overhead*). This state occurs when no thread is ready to be executed, thus NOPs have to be inserted at pipeline inputs instead.

Finally, if we compare the total number of clock cycles for execution of the L3 code on the LEON3 processor with the total number of clock cycles for execution of the UT code on the UTLEON3 processor, we can conclude that the latter one was 1.86 times faster on average. As in the previous case, it is mainly due to the elimination of both internal and external data dependencies.

7.1.2.2 FIR Filter with Different Unroll Factors

Several FIR filter executions with various unroll factors were compared in the second experiment. All versions are equivalent in terms of generated data outputs. A higher unroll factor should reduce the runtime in both the legacy mode and the microthreaded mode. In the case of the legacy mode the higher unroll factor enables a better arrangement of instructions as mentioned above. In the microthreaded mode higher unroll factors result in longer threads, which reduces the load on the thread scheduler.

The results are shown in Table 7.2 and Fig. 7.8. Examples with the unroll factor 1, 2 and 4 were used for both the legacy code (marked as L3 FIR 1x, L3 FIR 2x, L3 FIR 4x respectively) and the microthreaded code (marked as UT FIR 1x,

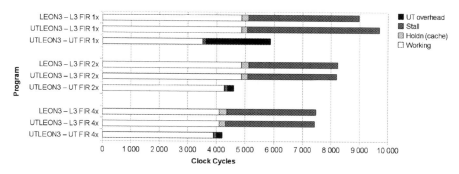

Fig. 7.8 Pipeline execution profile for a 26-tap FIR example

UT FIR 2x, UT FIR 4x respectively). The structure of the table and the graph is the same as in the previous experiment.

As expected, the number of pipeline stalls due to internal data dependencies (see the column *Stalls*) was smaller in versions with higher unroll factors. Nevertheless the reduction due to loop unrolling was noticeably smaller when compared to the reduction due to the use of microthreading. Considering the UTLEON3 processor, the application of microthreading reduces the number of stalls 35 times on average – while it represented 43 % of the overall runtime on average in the legacy code, it was just about 2 % of the overall runtime for the microthreaded code.

The higher unroll factor results in a lower number of stalls due to thread management (see the column *UT Overhead*). This overhead is 38 % for unroll factor 1 (UT FIR 1x), and decreases down to 5 % for higher unroll factors (UT FIR 2x, UT FIR 4x).

7.1.3 Summary

Sample DSP kernels have been coded in the common SPARC V8 assembler and in the extended version that supports microthreading. The program runs have been compared and analysed in terms of the overall runtime and pipeline execution profile. The results show that even a single-core implementation of a microthreaded processor can achieve better execution performance compared to the original architecture, achieving significantly higher instruction per cycle numbers.

7.2 Long-Latency Operations

Previous works [14, 19] have dealt with an analysis of multithreading efficiency with respect to the number of available hardware thread contexts (denoted N, see Table 7.3), average run length between cache faults (R), cache latency (L), register

Table 7.3 Parameters

Parameter	Description	(Units)
R	Average run length between two LL ops.	(cycles, instructions)
L_N	Non-pipelined latency, e.g. a cache fault	(cycles)
L_P	Pipelined latency, e.g. FP operation	(cycles)
N	Supported number of thread contexts	(number of threads)
B	Family blocksize	(number of threads)
c	Thread context size	(registers)
S	Context switch cost	(cycles)
f	Register file size	(registers)

LL long-latency, *FP* floating point

file size (F), and the required context size (the number of registers per thread; C). However, as the previous processor architectures did not have the concept of families of threads, they could not analyse the impact of the *blocksize* parameter (B) with its implications to fine-grain parallelism. In [14] only cache fault latencies (i.e. *non-pipelined* ones) were stochastically analysed, and the number of contexts (N) was considered a machine parameter, while in the UTLEON3 architecture it is a program-controlled (but implementation bounded) variable. This chapter discusses effects of this parameter on the efficiency of the microthreaded execution in the UTLEON3 processor. We show that the blocksize parameter has a similar effect as the number of thread contexts had in the previous architectures. The outcomes of the work can be used to improve the compilation technology for the microthreaded architecture.

7.2.1 Analysis of the Problem Situation

We classify long-latency (LL) operations in two groups: (a) *pipelined LL operations* are executed in a fully pipelined unit (e.g. an integer multiplier, a floating-point unit) which means the processor can initiate a new computation in the unit every clock cycle, and the processing takes L_P cycles. (b) *non-pipelined LL operations* must be served sequentially in the unit as they compete for some exclusive resource, e.g. a system/memory bus in the case of a cache fault, and we assume the operation takes L_N cycles to complete.

Upon creation of a family the total number of threads n in the family (equal to the number of iterations of a hypothetical *for*-loop) and the blocksize parameter B are specified by the program. The scheduler unit then reserves B entries in its thread table and allocates $B \cdot C$ registers in the (large, but shared) register file. Thus, in the family no more than B threads can execute concurrently at any moment. But if the requested number of resources cannot be allocated (for they were spent on existing k families: $\sum_k (B_k \cdot C_k) \leq F$), the hardware scheduler must reduce the value of B until the allocations succeed so that the computation can proceed.

a

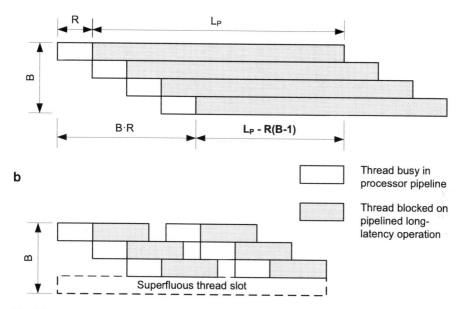

b

☐ Thread busy in processor pipeline

▨ Thread blocked on pipelined long-latency operation

Fig. 7.9 Thread schedule for pipelined long-latency operations. A new operation can be issued while the previous one is processing, e.g. a pipelined integer multiplication, a floating point operation. (**a**) **Linear region** $(L_P > R(B-1))$ – small blocksize; the CPU pipeline is waiting for the long-latency operation. (**b**) **Saturation** $(L_P < R(B-1))$ – large blocksize; the CPU pipeline is kept busy, but one family may consume excess resources

Clearly each family of threads contributes a different amount of parallelism to the processor execution. By optimizing the value of B for each family a compiler or assembly-level programmer can achieve better utilization of processor resources.

7.2.1.1 Impact of the Pipelined Long-Latency Operations

Let us assume a family of n threads with blocksize B. Each thread of the family executes $R-1$ short-latency (one cycle) instructions and one long-latency instruction; thus the run length between two long-latency operations is R cycles. The long-latency operation takes L_P cycles to complete, and it is fully pipelined. The thread switch cost is S cycles.

Figure 7.9 shows what happens in the processor pipeline with respect to the blocksize B and latency L_P. In Fig. 7.9a the long-latency operation has a latency L_P much higher than the combined computational load of B threads, i.e. $L_P > R(B-1)$. In this case the family operates in the *linear region* because increasing B gains more performance.

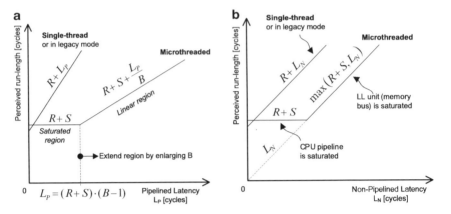

Fig. 7.10 Theoretical performance of long-latency operations. (**a**) Pipelined long-latency operations, e.g. a pipelined FP operation. (**b**) Non-pipelined long-latency operations, e.g. cache faults

In Fig. 7.9b the latency L_P is shorter than $R(B-1)$, and thus the program computation is limited only by the processor pipeline performance (and the context switch cost). Increasing the blocksize parameter B cannot gain more performance here because the pipeline is already *saturated*.

The perceived runtime in cycles per LL operation is plotted in Fig. 7.10a. As the total program runtime depends on the number of instructions executed in the processor, in practice we normalize it to obtain the execution efficiency to compare execution for different benchmarks. The efficiency of one family in the linear region $\eta_{P,lin}$ can be calculated by considering R to be the ideal number of *Busy* cycles:

$$\eta_{P,lin} = \frac{Busy}{Busy + Switching + Idle} = \frac{R}{R + S + \frac{L_P}{B}} = \frac{R \cdot B}{(R+S) \cdot B + L_P} \qquad (7.1)$$

In the saturation the efficiency does not depend on the blocksize, but the cost of context switches is more pronounced:

$$\eta_{sat} = \frac{R}{R+S} = \frac{1}{1 + \frac{S}{R}} \qquad (7.2)$$

Note that the saturated efficiency η_{sat} is the same for the pipelined and non-pipelined classes of long-latency operations.

7.2.1.2 Impact of the Non-pipelined Long-Latency Operations

A non-pipelined long-latency unit can process at most one request at a time because of an occupancy of some exclusive resource. Each request will take L_N cycles to process; any new request issued in the meantime must be blocked or stored in a FIFO queue to be served later. This is the model of a cache fault where each instance

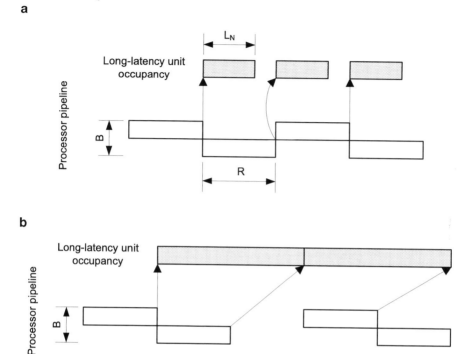

Fig. 7.11 Thread schedule for non-pipelined long-latency operations (e.g. cache faults). A new operation can be issued on completion of the previous operation. (**a**) $L_N < R$ – the processor pipeline is saturated. (**b**) $L_N > R$ – the long-latency unit (memory bus) is saturated

requires an access to the memory bus. An impact of a limited memory bandwidth on a multithreaded processor was theoretically studied e.g. in [6].

The best case in the microthreaded mode occurs when the processor pipeline or the long-latency unit are saturated (Fig. 7.11). Blocksize $B = 2$ is theoretically sufficient to keep at least one of the two resources fully occupied so that the program perceives minimal latency $max(R + S, L_N)$ (Fig. 7.10b). Thus in the microthreaded mode for non-pipelined long-latency operations the efficiency η_N of a family of threads does not depend on the blocksize:

$$\eta_N = \frac{R}{max(R + S, L_N)} \tag{7.3}$$

7.2.2   Experimental Evaluation

Experiments were carried out on an FPGA-synthesizable VHDL model of the UTLEON3 processor. We modified the processor's integer multiplier to simulate a long-latency pipelined unit with an arbitrary delay of L_P cycles. Similarly, the model of the main memory was extended so that it can simulate a non-pipelined delay of L_N cycles. A synthetic benchmark program creates a family of threads in which each thread executes $(R - 1)$ short-latency (one cycle) and one long-latency instructions. The family blocksize value is specified by the setblock instruction before a family is created. Efficiency of early voluntary context switches is compared by using the swch modifier in the first (short-latency) instruction that depends on the long-latency one and comparing it to a run without the swch modifier.

The results in Fig. 7.12a show CPU utilization efficiency η_P with respect to the pipelined long-latency operation L_P that was simulated by the modified multiplier unit. The run-length of the test program was $R_P = 15$ cycles. Different blocksizes $B_P = \{6, 8, 16, 24\}$ and the presence of the swch modifier were evaluated ($S_{without-swch} = 3$ cycles, $S_{with-swch} = 1$ cycle).

As shown in the analysis the blocksize parameter B determines the point (latency L_P) when the family transitions from the saturated region, where it executes with the maximal efficiency η_{sat} (Eq. 7.2), to the linear region, where the efficiency $\eta_{P,lin}$ (Eq. 7.1) decreases with L_P. Presence of the swch modifier affects the context switch cost S that influences the saturated efficiency $\eta_{sat} = \frac{R}{R+S}$ (Eq. 7.2). Without the swch modifier the theoretical saturated efficiency with the given $R = 15$ is 0.83, but with the swch modifier it is 0.93. This analytical prediction agrees with the measurement in Fig. 7.12a.

Figure 7.12b shows the CPU utilization efficiency η_N with respect to a non-pipelined long-latency operation L_N. In this case the run-length of the test program was $R_N = 49$ cycles, and different blocksizes $B_N = \{4, 8, 16\}$ and the presence of the swch modifier were evaluated.

As expected the blocksize parameter does not influence the CPU efficiency η_N so much when compared to the previous case. The effect of the swch modifier is not very pronounced in the plot, but that is because $R_N \gg S$ in the presented measurement.

We believe that the chosen values have a connection to real-world situations. In the pipelined long-latency operation the $R_P = 15$ cycles corresponds to a 6 % arithmetic intensity. For example the *Discrete Cosine Transform* (DCT) kernel from the *Independent JPEG Group* [8] compiled for the SPARC architecture has an arithmetic intensity of 15 %. In the non-pipelined long-latency operation the $R_N = 49$ cycles corresponds to a cache miss rate of 2 % over all instructions executed; given a typical ratio of memory access instructions of 20 % in common programs (every fifth instruction is Load/Store), this implies a 10 % D-Cache miss rate. Albeit quite large for contemporary single-threaded processors, this miss rate can be expected in processors that execute multithreaded workloads that increase pressure on the cache subsystem.

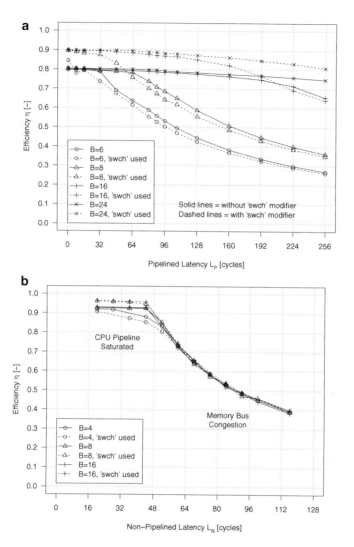

Fig. 7.12 Influence of the blocksize parameter on execution efficiency – experimental results obtained for the FPGA implementation of the UTLEON3 processor. (**a**) *Efficiency of pipelined long latency operations.* For short latencies the efficiency is optimal even with small blocksize values (the processor is saturated) and the effect of the 'swch' modifier is most profound. Longer latencies require higher blocksize values to stay in the saturation. Measured for $R_P = 15$ cycles, blocksize $B_P = \{6, 8, 16, 24\}$. (**b**) *Efficiency of non-pipelined long latency operations.* Once the memory bus is congested increasing the blocksize value does not improve the execution efficiency. Measured for $R_N = 49$ cycles, blocksize $B_N = \{4, 8, 16\}$

7.2.3 Summary

We have analysed the impact of long-latency instructions, the family blocksize parameter, and the thread switch modifier on execution efficiency of families of threads in the UTLEON3 processor that implements the microthreaded model. Experimental evaluations run in the FPGA implementation of the processor support the analysis.

We classify long-latency operations as either *pipelined* or *non-pipelined*. The analysis shows that the blocksize parameter that controls thread resource allocation in the processor hardware has profound effects when the latency is *pipelined*, i.e. increasing the blocksize value can improve the performance up to the saturation point η_{sat}. In the *non-pipelined* long-latency case the efficiency reaches its maximum even for small blocksize values, beyond which it cannot improve due to the occupancy of an exclusive resource (memory bus congestion).

As the compiler specifies the blocksize parameter for each family of threads individually, the analysis can be used to optimize resource usage. The compiler should specify smaller blocksize values for memory-intensive families of threads to save resources, while compute-intensive families with many pipelined FP operations could benefit from larger blocksize values. Ideally each family should operate just at its saturation point where the efficiency reaches its maximum while the resource utilization is optimal.

The presented analysis of the impact of the non-pipelined long-latency operations in UTLEON3 relates to the current implementation of its memory subsystem. The memory is connected over the AMBA AHB bus which cannot serve multiple memory transactions concurrently, hence the memory is a non-pipelined resource. More advanced memory buses (such as AMBA AXI) can process several outstanding requests in an interleaved manner. An improved interconnect architecture and a more advanced memory controller could be characterized (to some extent) as a pipelined resource from the processor point of view, which performance-wise would enable better results.

Chapter 8
Hardware Families of Threads

This chapter describes *Hardware Families of Threads*: a method of connecting hardware accelerators (co-processors) to a microthreaded processor.

8.1 Interaction of Software Families and Hardware Families

The UTLEON3 processor is backwards binary compatible with LEON3: after reset it starts in the legacy mode. To switch to the microthreaded mode the instruction `launch` has to be executed. Afterwards the other microthreaded instructions (`allocate`, `set...`, `create`, etc.) can be used as depicted in Fig. 4.2. Figure 8.1 shows a block diagram of the UTLEON3 processor extended with a new hardware accelerator for families of threads on the right-hand side of the picture, and Fig. 8.2 illustrates the principle of a HW family execution.

Hardware families of threads are designed to have no impact on the instruction code of a microthreaded program. Hardware and software families that perform an *identical task* are coupled together by the memory address of the software family. In software this is set by the `setthread` instruction as the memory address of the first instruction of the thread code. At the same time it serves as a *tag* that uniquely identifies a hardware family that implements the same functionality in the hardware accelerator. The subsystem of the hardware families keeps an associative *Thread Mapping Table* that maps software memory addresses to particular instances of hardware accelerators and their configurations. A system code must execute a subprogram (in the legacy mode) to initialise the subsystem of hardware families. The key part of the initialization process is writing the family look-up addresses to the associative table; see the arrow marked (1) in Fig. 8.2.

After entering the microthreaded mode the processor executes all instructions as usual up to the `create` instruction. When the `create` instruction is issued, the processor checks the *Thread Mapping Table*; see Fig. 8.2. In the example the address of the software thread 0x200 is looked-up in the table. If the address is found, the system will execute the family in the hardware accelerator. Conversely,

M. Daněk et al., *UTLEON3: Exploring Fine-Grain Multi-Threading in FPGAs*,
DOI 10.1007/978-1-4614-2410-9_8, © Springer Science+Business Media, LLC 2013

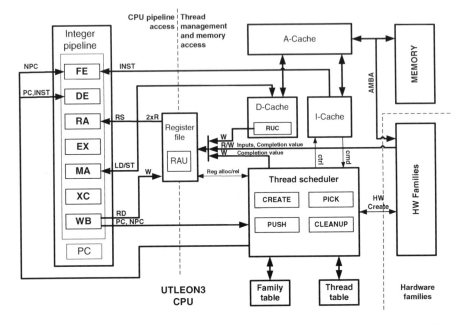

Fig. 8.1 Block diagram of the UTLEON3 processor with a hardware accelerator for families of threads

if the address is not found, the system will create an ordinary software family in the processor. From an application point of view the creation scheme is transparent.

8.2   Analysis of Feasibility of Implementing a Family in Hardware

Before deciding that a given family should execute in hardware one must analyze the possible performance advantage. In general not all families are suitable to be implemented in hardware. For example, threads executed only a few times (a small family) or those that have a complex behaviour may not present an opportunity for a speed-up. Profiling techniques should be used to identify families suitable for hardware acceleration. Good candidates are families that exhibit a simple data-flow behaviour without many data dependencies. In these cases a pipelined implementation of the hardware family should achieve a rate of one result per clock cycle.

Achieving the rate of one result per cycle is easy when there are no data dependencies between threads. Figure 8.3 illustrates the situation for a hardware thread that has four pipeline stages (A–D). At each clock cycle a new thread can be started (T0-T2) and – after the initial latency of $L = 4$ cycles–one result per clock cycle is obtained.

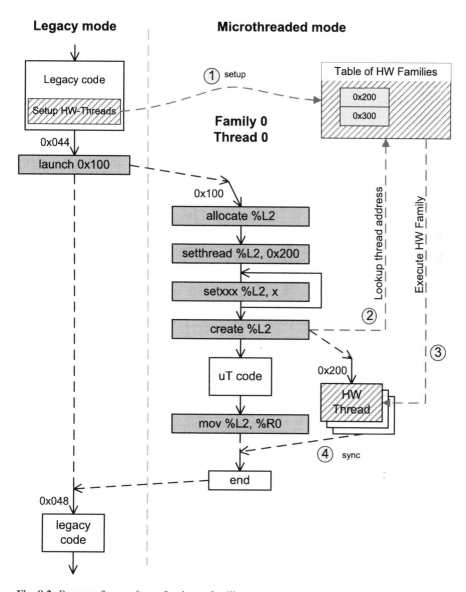

Fig. 8.2 Program flow: software/hardware families

Fig. 8.3 A thread pipeline without dependencies

In the presence of data dependencies the situation is more complicated. We will consider threads with an associative coupling operator. The computation of the family of threads can be described by the formula

$$d_0 \circ T_0 \circ T_1 \circ T_2 \circ \ldots \circ T_n = s_n$$

where $\circ$ is an associative coupling operator, $T_0 - T_n$ are threads, d_0 is a dependent register of thread T_0 (mapped to the parent thread), and s_n is a shared register of the last thread T_n.

Assuming the threads are associative, we can regroup the threads (add brackets) as needed:

$$(d_0 \circ T_0 \circ T_1 \circ \ldots \circ T_{k-1}) \circ (e \circ T_k \circ T_{k+1} \circ \ldots \circ T_{2k-1})$$

$$\circ (e \circ T_{2k} \circ \ldots \circ T_{3k-1}) \circ (e \circ T_{3k} \circ \ldots \circ T_{4k-1}) = s_n$$

In the example we have partitioned the family (of size $n + 1$) into four groups of k threads:

$$k = \frac{n+1}{L} = \frac{n+1}{4}$$

as the latency of one thread is still $L = 4$ clock cycles. We had to introduce a neutral element e into the formula. The neutral element is used as the input value (a dependent) in all groups except the first one. The numerical value of the neutral element depends on the nature of the associative operation (e.g. for addition it is 0, for multiplication it is 1).

Having partitioned the threads into the groups it is easy to construct a pipeline schedule, because the computation in the distinct groups is independent – see Fig. 8.4. We simply interleave the computation of distinct groups one after another, so by the time the second thread of a group is starting, the previous thread in the group has just finished. Finally we need to perform reduction of the partial results at the end of the computation (denoted R in Fig. 8.4) to obtain the final result s_n.

In practical programs the coupling operator $\circ$ usually is not only associative, but also commutative. In such cases we can reorder the threads and use a different partitioning:

$$(d_0 \circ T_0 \circ T_4 \circ \ldots \circ T_{4k-4}) \circ (e \circ T_1 \circ T_5 \circ \ldots \circ T_{4k-3})$$

$$\circ (e \circ T_2 \circ \ldots \circ T_{4k-2}) \circ (e \circ T_3 \circ \ldots \circ T_{4k-1}) = s_n$$

Again the threads were partitioned in four groups. The first group consists of threads that satisfy: $index(T) \equiv 0 \, (mod \, 4)$, the second group: $index(T) = 1 \, (mod \, 4)$, and so on. The advantage of this partitioning scheme is that the pipeline schedule (after interleaving the threads of different groups) will issue threads in the natural order: $T_0, T_1, T_2, T_3, T_4, \ldots, T_n$ (apart from that the schedule looks the same as in

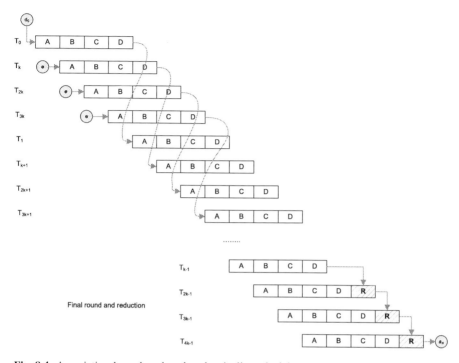

Fig. 8.4 Associative dependent threads – the pipeline schedule

Fig. 8.4). This is a very important property for it is more likely to provide better memory access pattern along with a simplified hardware implementation.

To conclude, it was shown that we can provide a well-performing pipelined implementation of data-flow threads in hardware not only when the threads are independent, but also in the case of dependent threads with commutative and associative coupling.

8.3   Implementing a Family in Hardware

As explained above, we will focus on a class of threads with a dataflow behaviour. We will require threads to be independent, or to exhibit an associative and commutative behaviour.

Each hardware family class is unique and exhibits different properties. To deal with the complexity, the family-specific pipelines are encapsulated in the *Basic Computing Element* cores (BCE, see Fig. 8.5). The BCE has a unified I/O interface that consists of dual-port memories (A, B, Z in the picture), common design primitives found in modern FPGA platforms. Although the actual number of memories per one BCE may vary, the structure of the BCE interface remains stable.

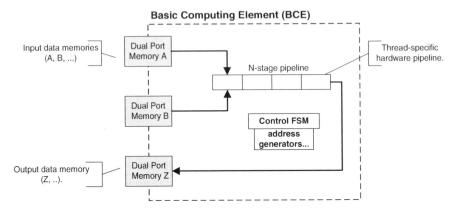

Fig. 8.5 Encapsulation of a thread in a BCE

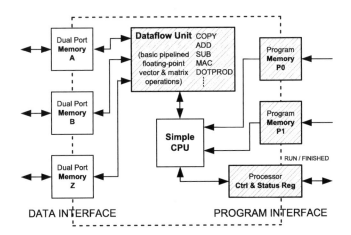

Fig. 8.6 A generalized BCE core

This in turn simplifies the overall design of the proposed HW Family subsystem described in the following text.

To ease partial reconfiguration in the FPGA and facilitate hardware reuse the internal structure of some BCE cores may be further generalized as dataflow/control units as depicted in Fig. 8.6. The generalized BCE is more complex, but it can be reconfigured by changing its firmware in its program memories (P0, P1).

8.4   Example: Developing a New Hardware Family

The implementation methodology for creation of hardware families will be illustrated on an simple FIR filter. A FIR filter output is given by

$$z_k = \sum_{i=0}^{L-1} b_i x_{k+i} \tag{8.1}$$

```
int x[N], z[N], b[L];

for (k = 0; k < N-L; k++) {
  int s = 0;
  for (i = 0; i < L; i++) {
    s += b[i] * x[k+i];
  }
  z[k] = s;
}
```

```
int x[N], z[N], b[L];

F1: create_family (index int k;
                   start=0; limit=N-L-1) {
  int s = 0;
  F2: create_family (index int i;
                     start=0; limit=L-1
                     global={k}; shared={s}) {
    s += b[i] * x[k+i];
  }
  z[k] = s;
}
```

Fig. 8.7 A complete FIR program; legacy (left), microthreaded (right)

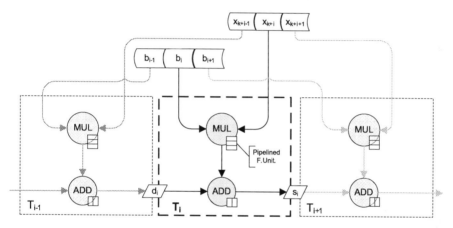

Fig. 8.8 FIR: Pipelined computation in inner threads

where x and z represent filter inputs and outputs, respectively. The parameter L specifies the length of the filter. In typical embedded DSP applications (echo cancellation, ADSL) the length is often of the order of tens of elements (taps); we can safely assume here a typical $L = 32$.

Figure 8.7 shows a complete software implementation of a FIR filter that computes the filter equation over an array of N elements. The inner family F2 implements the filter equation by unrolling the sum into a family of dependent threads (using a shared integer s). These threads are coupled by the addition operator (line s += b[i] * x[k+i];) which is both associative and commutative. Therefore the inner family of threads can be scheduled in an optimal way suitable for efficient hardware implementation (Fig. 8.8).

Another possibility is to partition the program one level higher, i.e. to implement the family F1 in hardware. This has two advantages. First, threads of the family are independent (no shared registers). Second, the family F1 offers a better granularity of the computation (because $N \gg L$). Therefore, we will transform family F1 (and everything below it) to a hardware family.

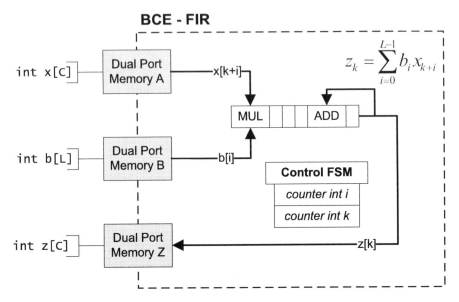

Fig. 8.9 A version of a BCE that implements FIR

Figure 8.9 shows a possible implementation of the FIR family in the BCE core. The dual-port memory A contains an input array x[], the memory B holds an array of the FIR coefficients b[], and finally the memory Z stores the resulting array z[].

The picture suggests that this particular specialized BCE can compute only the FIR family, but this does not have to be so. We could use the generalized BCE (Fig. 8.6), and the computation of the FIR family will be coded in the BCE firmware as a sequence of basic kernels to be applied on the data to generate the desired result. The interface (dual-port memories) remains the same for different internal implementations.

8.5   Conceptual Structure of the Hardware Family Subsystem

The previous sections in this chapter have described a conversion of a particular software family of threads to a form suitable for accelerated hardware. The hardware families are implemented and executed in the *BCE* cores that have unified interfaces based on dual-port memories (Block RAMs in the FPGA). In this section we will describe the control subsystem for the hardware families (*HWFAM*). The conceptual block diagram is shown in Fig. 8.10.

The hardware families are encapsulated in the BCE cores; two instances are shown in Fig. 8.10 at the top. Depending on its complexity a BCE core can be

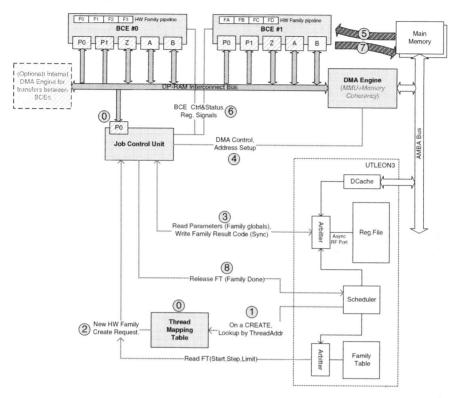

Fig. 8.10 A conceptual block diagram of the hardware family subsystem

tailored to contain only a single-function or it can be programmable. The BCE interface, consisting of DP-RAM memories P0, P1, Z, A, B, is connected to an internal interconnect bus. The bus is driven by the *DMA Engine*. The DMA engine transfers data between the BCE DP-RAM memories and an external main memory over the system AMBA bus. The DMA engine should implement a bus coherency protocol to allow seamless data sharing with the processor cache subsystem. The *Job Control Unit* (JCU) is the central control machine of the subsystem. Its function can be summarized as follows: after detecting a HW-Family *create event* DMA transfers are initiated, then BCE executes its function, and finally family completion is signalled back to the scheduler (a detailed scenario will be given below). The JCU cooperates with the *Thread Mapping Table* unit. This unit functions as a filter that determines which software *create*s (i.e. the processor's create instructions) are to be handed-over to the JCU to be executed in the hardware accelerator.

8.6   Elementary Use-Case Scenario

This scenario gives a basic example how a family can be executed in the hardware family subsystem. The situation is illustrated on a family that implements a *Finite Impulse Response* (FIR) filter (see Sect. 8.4). A particular software implementation of the algorithm has the following calling convention:

Register	Description
%tg0	Pointer to an input vector X
%tg1	Pointer to an output vector Z
%tg2	Pointer to a vector of coefficients B
%tg3	Number of coefficients (elements in vector B; filter taps)
Family start/limit	Generates indices over the input and output vectors X and Z

Listing 10 shows one possible instantiation of the FIR family in an application. When the processor issues the `create` instruction at line 15 in the listing, the following sequence of events happens:

1. The scheduler sends a query to the *Thread Mapping Table* unit, checking it whether the family of threads being created could be run in the hardware accelerator (Fig. 8.10, step (1)). The query includes:

   - *Thread address* – identifies the family of threads (the address used in the `setthread` instruction);
   - *Synchronizing register number* – the register where the family exit code is stored on completion (this is taken from the `create` instruction);
   - *Family ID* (FID) – it is used to release the line in the family table when the computation finishes.
   - *Base addresses of the Global and Shared registers* so that the family parameters can be retrieved.
   - *Start/step/limit* parameters of the family.

2. The *Thread Mapping Table* receives a query from the scheduler. The thread address is treated as an opaque tag that uniquely identifies the hardware family. The tag is looked up in the table to find out if the family can be executed in the hardware accelerator. The table also specifies the minimum and maximum size of the family that can be executed in the hardware accelerator.

3. The *Thread Mapping Table* communicates its result back to the scheduler:

   (a) The result is positive, meaning the family can be executed in hardware. The scheduler ceases family creation, i.e. it just marks the family as allocated. At the same time the *Thread Mapping Table* passes the parameters of the accepted new family to the *Job Control Unit*.

   (b) The result is negative, meaning the family cannot be executed in the hardware accelerator subsystem (the hardware configuration does not exist), or

---

Listing 10 Creating the FIR family (mtsparc assembler listing)

```
      set  xvec,  %tl0
2              /* %tl0 = address of xvec; becomes %tg0 in the family */
      set  zvec,  %tl1
4              /* %tl1 = address of zvec; becomes %tg1 */
      set  bvec,  %tl2
6              /* %tl2 = address of bvec; becomes %tg2 */
      set  (BVEC_LEN−1)*4,  %tl3
8              /* %tl3 = bvec_len; becomes %tg3 */

10    /* FIR family */
      allocate %tl20
12             /* allocate a new family of threads */
      setstart %tl20, 0
14             /* set index start = 0*/
      set  (XVEC_LEN−1)*4,  %tl21
16    setlimit %tl20, %tl21
               /* set index limit = length of X array */
18    setstep %tl20, 4
               /* set index increment = 4 */
20    set  fir_family, %tl21
      setthread %tl20, %tl21
22             /* set thread function */
      setblock %tl20, BLOCKSIZE_1
24             /* set the blocksize parameter */
      create %tl20, %tl20
26             /* run the family */
      mov %tl20, %tl21
28             /* wait for the family to terminate */
```

---

there are not enough resources left in the accelerator to execute the family. The scheduler continues with creating and executing the family in software as usual.

4. The *Job Control Unit* (JCU) receives a request to run a new family (step (2) in Fig. 8.10). The request consists of identification of the family together with the register number to store the exit code to and the FID. The JCU writes the data into its internal table of running hardware families.
5. The JCU assigns BCEs to the new job, determines the BCE firmware programs that are then downloaded to the selected BCEs (if needed) (see the 'BCE Setup Phase' in Fig. 8.11). The firmware is loaded from the main memory through the DMA engine into the P0 or P1 BRAM of the BCE. The JCU can also select a BCE that has a firmware already loaded from the previous run. Thus the two distinct program memories (P0, P1) in the BCE can act like a firmware cache.
6. In the meantime, the JCU reads the family global and shared registers through the asynchronous register file port of the UTLEON3 processor. The number of the fetched registers is specific to each hardware family, and it is part of the

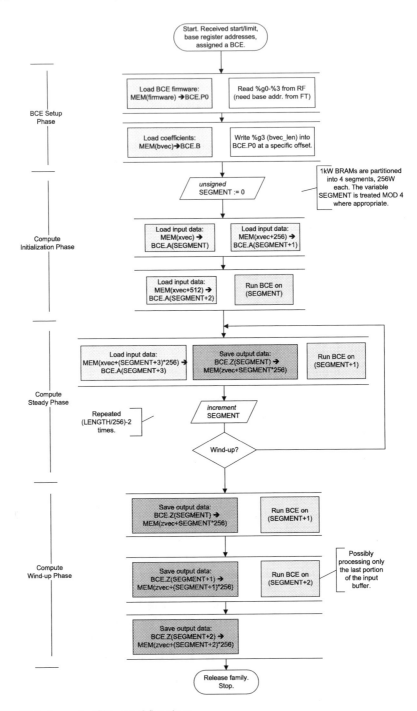

Fig. 8.11 A sample JCU control flowchart

JCU programming. In the FIR example, there are four global registers to be read from the RF: %tg0–%tg3.

7. In the case of the FIR example, the JCU initially schedules one DMA operation to transfer the vector of coefficients (bvec = %tg2, length %tg3) from the memory to BRAM B of the assigned BCE (step (5) in Fig. 8.11).

8. The total number of words to compute in the accelerator is: $LENGTH := (LIMIT - START)/sizeof(word)$. The work is divided into batches of no more than 256 words as in our particular example the BCEs have only 1,024-word local memories: $BATCHES := LENGTH/256$. Thus each BCE data BRAM consists of 4 segments of 256 words each.

9. The computation schedule in the BCE is organized in three phases: initialization, steady state and wind-up. The BCE program is run in iterations, each iteration consumes one segment (256 W) of the input data in BRAM A, and produces one segment of results in BRAM Z. Steps (4)–(7) in the Fig. 8.11 are executed repeatedly.

   (a) During the *initialization phase* the first three segments (i.e. the first 768 W) of BRAM A are filled with the FIR input data (buffer at address %tg0). Then, the BCE is run once to process the first segment.

   (b) In the *steady state phase*, in which the computation spends the majority of the time, three operations are executed in parallel during each iteration. An iteration is completed when all the three operations scheduled have completed:

      • Load the input data in the next input segment of BRAM A. Segment numbers (#0 to #3) are treated modulo 4. (In the very first iteration after the initialization phase the last segment (#3) of BRAM A is loaded.)
      • Save the output data from the last output segment of BRAM Z. (In the very first iteration after the initialization phase the first segment (#0) of BRAM Z is saved.)
      • Run the BCE once, configure it to process the segment already prepared in BRAM A. (In the very first iteration after the initialization phase segment #1 from BRAM A is processed, and the results are written to segment #1 in BRAM Z.) Note that when the BCE is processing an input segment n, the segment $(n + 1)$ is already loaded in the input BRAM A. This is to allow the BCE algorithm to see the next data in the input stream continuously.

   (c) During the *wind-up phase* we let the BCE process all the remaining data. There can be a partial DMA transfer (not the whole 256-word segment), and the BCE can be configured to process only a part of the input buffer (LENGTH mod 256).

10. When the data processing has completed, the JCU writes the family exit code to the family synchronization register. The JCU also signals the scheduler to release the family (step (8) in Fig. 8.11).

8.7   Sample Design and Its Implementation

The UTLEON3 processor with the *Hardware Families of Threads* was implemented in the Xilinx XUP-V5 board. Figure 8.12 shows the block diagram of the implemented system. The following modules are depicted from the top-left to the bottom:

- *BCE Worker*: a very simple BCE core that computes $Z_i := A_i + B_i$ was implemented to develop the BCE data/control interfaces.
- *XBAR Switches*: a simple switched bus was implemented to facilitate data and control transfers within the subsystem. The switch component is instantiated twice: the *BRAM XBAR* carries data between the local dual-port memory banks ('BRAMs') and the DMA Engine; the *CONF XBAR* is used for configuration access to the modules.
- *DMA Engine*: a controller that transfers data between the main memory (over the system AMBA bus) and the local dual-port memory banks ('BRAMs').
- *Host Bridge*: it maps the local dual-port memories ('BRAMs') and the module configuration memories to the AMBA address space. This allows an external entity, such as a legacy setup code running in the UTLEON3, to access the internals of the subsystem easily.
- *UTLEON3 Interface*: through this interface the *create* queries from the microthreaded scheduler are issued, and the processor register file can be accessed.
- *Job Control Unit*: the controller responsible for running a family in a particular BCE core and managing all data transfers via the DMA Engine.

The following text describes a detailed implementation of the HW-Families subsystem.

8.7.1   *Internal Backbone Buses and the Host Bridge*

The two XBAR switches create a backbone to which all the modules are connected. They facilitate data and control transfers within the subsystem. The switches have a configurable number of master and slave ports (see Fig. 8.13).

Masters arbitrate a path through the switch by setting the appropriate slave address (xbar_master_down.slave_addr), and activating their request signal (xbar_master_down.request). In the next clock cycle the xbar signals if an access was granted (xbar_master_up.granted). If not, the master continues holding the request line active until it eventually succeeds.

When the access has been granted, the master sends data through the switch (xbar_master_down.data_dwn, xbar_master_up.data_up). During the granted phase a master must not change the target slave address (a master must not change the target slave address while holding the request line active). Once a path has been granted to a master, it will stay granted to the master as long as the master is holding its request line active. This simplifies the design of both the master

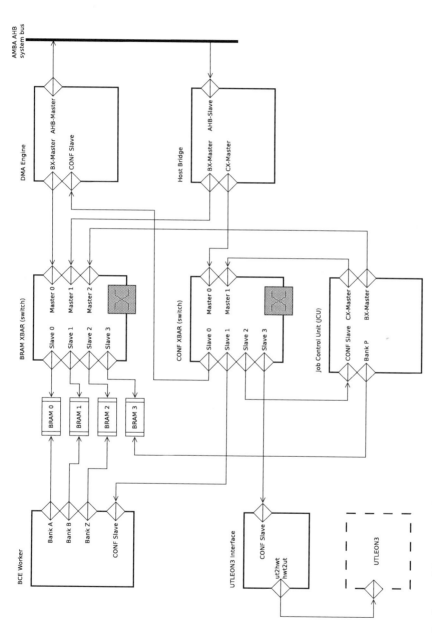

Fig. 8.12 Implemented demonstrator design (block diagram)

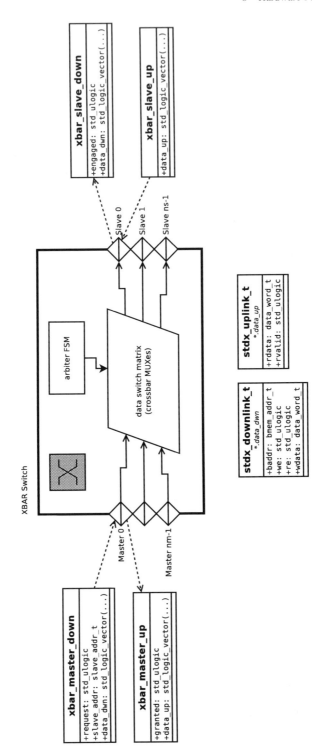

Fig. 8.13 Internal structure of the XBAR switch and its interface

Table 8.1 BRAM and CONF XBAR slave devices and their AMBA address space mapping. In the demonstrator design: $xxx = 0xB00$, and $yyy = 0xB01$

AMBA address space extent	Slave	Function
xxx0 0000 – xxx0 0FFF	BRAM 0	Bank A of the BCE core
xxx1 0000 – xxx1 0FFF	BRAM 1	Bank B of the BCE core
xxx2 0000 – xxx2 0FFF	BRAM 2	Bank Z of the BCE core
xxx3 0000 – xxx3 0FFF	BRAM 3	Bank P of JCU: PicoBlaze firmware
yyy0 0000 – yyy0 03FF	CONF 0	DMA Engine configuration registers
yyy1 0000 – yyy1 03FF	CONF 1	BCE core configuration registers
yyy2 0000 – yyy2 03FF	CONF 2	JCU configuration registers
yyy3 0000 – yyy3 03FF	CONF 3	UTLEON3/HWT interface configuration registers

and the slave device. Only by de-activating the request line the master relinquishes the path, and the granted signal is deactivated by the xbar in the very next clock cycle. To change a target slave a master must relinquish its current path (deactivate request), set a new target slave address, and then it may re-activate the request line to arbitrate a path.

As depicted in Fig. 8.12, the 'BRAM XBAR' is to carry data between the local dual-port memory banks (BRAMs 0–3 connected as Slaves 0–3) and the DMA Engine (Master port 0). The local memory banks can be also accessed from the Job Control Unit (through the port Master 2) and the Host Bridge (through the port Master 1).

The 'CONF XBAR' switch is a configuration access bus. It provides a unified protocol for accessing the control and status information stored in all the other units from the Host Bridge and the Job Control Unit. Each of these slave units provides a 256-word configuration address space for its registers.

Table 8.1 displays the device numbers of the slaves connected to the switches, along with their AMBA address space mapping. The transparent AMBA mapping of the configuration registers and the local memory banks is performed by the *Host Bridge* module. The bridge is an AMBA-AHB slave device connected to the system bus. It implements two continuous address space extents, prefixed *xxx* and *yyy* in Table 8.1, to which the XBAR slave address spaces are mapped. In the demo design the prefixes are assigned as follows:

- $xxx = 0xB00$, and
- $yyy = 0xB01$.

8.7.2 DMA Engine

The DMA Engine autonomously transfers data to and from the main memory (over the system AMBA bus) and the local dual-port memory banks ('BRAMs 0–3'). It implements 16 independent channels (configurable during core instantiation) that can upload or download chunks of data to the local memory banks.

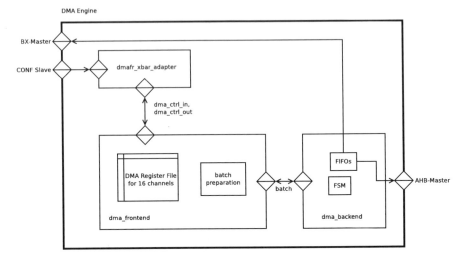

Fig. 8.14 The DMA engine (block diagram)

The DMA engine internal structure is depicted in Fig. 8.14. It has three main interfaces:

- *BX-Master* is connected to the BRAM XBAR Master port 0. Through this port the data is transferred to and from the local memories.
- *AHB-Master* is an AMBA-AHB Master port. This interface is used to access the AMBA system bus and the external system memory.
- *CONF Slave* is a slave port of the CONF XBAR. Through this interface the configuration registers are accessed.

The engine is configurable through the 'CONF Slave' interface. Its internal registers are mapped to the 256-word address space available through the interface, see Fig. 8.15. Each of the 16 Channels has four configuration registers as depicted in the picture. The purpose of the registers is further detailed in Table 8.2. Once the channel registers are set, the channel can be enqueued for processing. This is done by writing the channel number into the 'Channel FIFO' register in the 'Global Ctrl. Reg.' area. A bit-map of the all currently active (enqueued) channels is also provided along with the FIFO status bits ('A', 'B' in the figure).

8.7.3 BCE Core

A simple BCE core that implements a family of threads $Z_i := A_i + B_i$ was developed for demonstration purposes. The core is connected to three dual-port memory banks ('BRAM 0–2'). The first two contain the input data. When the computation is over, the third one contains the results.

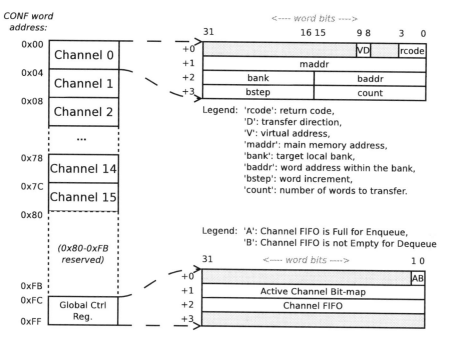

Fig. 8.15   The DMA engine: configuration address space

Table 8.2   DMA channel registers

Register	Data type	Function
direction	std_ulogic (1b)	Direction of the data transfer; 0=download into BRAM, 1=upload to the main memory
count	count_word_t (up to 16b)	Number of words to transfer; unsigned
maddr	memory_addr_t (32b)	Memory base address of the transfer
bank	slave_addr_t (up to 16b)	Identification of the target BCE BRAM, i.e. the slave target number
baddr	bmem_addr_t (up to 16b)	Starting word address within the target BRAM
bstep	count_word_t (up to 16b)	*baddr* increment; signed
va	std_ulogic (1b)	*maddr* is virtual; At present this has to be zero as this feature has not been implemented
rcode	dma_result_code_t (4b)	Result of the transfer: "0000" means ok, "0001" means memory bus error

The core is controlled through the CONF interface. The configuration address space, depicted in Fig. 8.16, is divided in two regions: the *input region* (words 0x00–0x7F) and the *output region* (words 0x80–0xFF). The core is controlled through the *Control* and *Status Words*. The other words in the configuration space are used to pass additional parameters such as the number of words to be computed and the starting addresses.

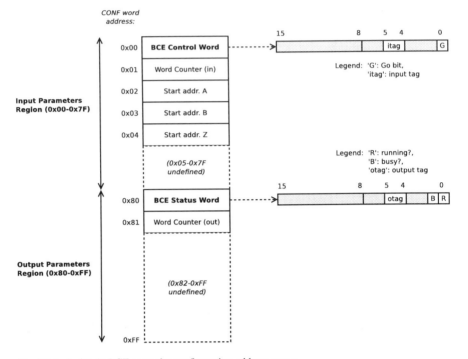

Fig. 8.16 A simple BCE core: the configuration address space

The input *Control Word* contains the 'Go' bit (G, 1b) and the *input tag* (itag, 2b). The output *Status Word* contains the 'Running' bit (R, 1b), the 'Busy' bit (B, 1b), and the *output tag* (otag, 2b). The 'G' bit is used as a general enable/disable; when cleared, the core is reset. The 'R' bit indicates if the core is running; basically this status bit reflects the input 'G' bit. The 'B' bit indicates if the core is busy, i.e. whether it is performing a computation. Figure 8.17 displays the state diagram of a BCE core.

The *itag* and *otag* fields are used to implement a barrier synchronization when submitting tasks. Whenever *itag* ≠ *otag*, the core is busy performing the requested computation. Once the work is done, it sets *otag* := *itag*. Then the user can retrieve the results, set up new parameters, and trigger a new computation by writing an *itag* value different from the previous one.

8.7.4 UTLEON3 Processor Interface

The *UTLEON3 Processor Interface* unit implements a direct interface device between the UTLEON3 scheduler and the HW-Families subsystem (the indirect interface is via the *Host Bridge*). Figure 8.18 shows its block diagram.

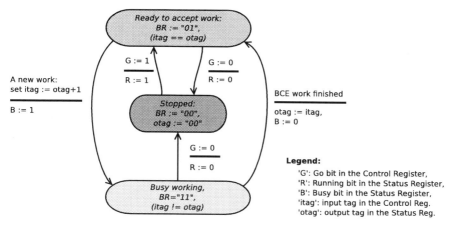

Fig. 8.17 The BCE state diagram

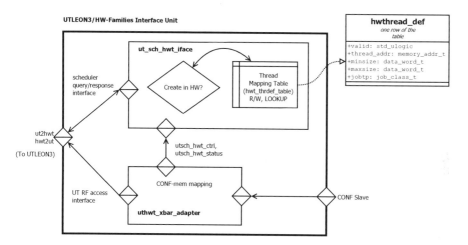

Fig. 8.18 The UTLEON3/HWT interface unit (block diagram)

The unit implements the UTLEON3 scheduler query/response interface. Its main part is the *Thread Mapping Table* (register file) that contains the specification of the hardware families of threads available in the particular implementation. The unit uses this table to decide, upon a family-create query from the UTLEON3 scheduler, whether a particular family of threads can be created in the hardware accelerator.

The structure of one row of the *Thread Mapping Table* is visible in Fig. 8.18. A principal component of the table is the *thread_addr* field. This field's value is matched against the software thread address given by the setthread instruction to pair the hardware and software implementation of the same family of threads. Furthermore, the *minsize* and *maxsize* fields can be used to filter out families of threads whose size (*limit* − *start*) is too small (or too big) to be executed in the hardware accelerator with good efficiency.

Figure 8.19 depicts the layout of the configuration address space of the device. The address space is divided in four regions:

1. *Thread Mapping Table* at words 0x00–0x7F. The table can contain up to 16 entries.
2. *UTLEON3 Register File Interface* at words 0xD0–0xDF. The interface allows the JCU to directly read/write the processor registers through its asynchronous port. This is useful for reading family global registers and for writing the family synchronization register.
3. *UTLEON3 Scheduler Interface* at words 0xE0–0xEF. When a family *create* from the processor scheduler has been accepted, it is visible in this area for the JCU to read and interpret. When the create message has been consumed by the JCU, it can be dequeued by setting bit 'D' in the Control Register (word 0xF0).
4. *Ctrl+Release Registers* at words 0xF0 and 0xF1. The Control Register allows to enable/disable the interface unit ('E' bit), and to dequeue the actual create message from the Scheduler Interface area. The *FID Release* field allows to send the *Family Release* response back to the UTLEON3 processor scheduler when a particular family of threads has been completed in the hardware.

8.7.5 Job Control Unit

The *Job Control Unit* is a controller responsible for interpreting a family create message, which was received from the UTLEON3 scheduler, driving the DMA Engine, and controlling the BCE cores. The JCU implementation can be either programmable, employing a simple processor, or a hard-wired with a fixed finite state automaton (FSM). As a part of the development process three solutions have been implemented:

1. *JCU-FSM*: a hard-wired FSM implementation. This JCU version is not programmable, but it is the fastest one. It gives us the lower bound on the control overhead required to execute a family in the hardware accelerator.
2. *JCU-PB8*: uses a simple programmable 8-bit CPU *PacoBlaze 3* (equivalent to the KCPSM3 from Xilinx). It was discovered that the *PicoBlaze 3* CPU is not very suitable for the function, mainly due to its 8b design and unfitting I/O port interface.
3. *JCU-PB16*: uses a programmable 16-bit CPU *PacoBlaze 3e*, loosely based on the KCPSM3 from Xilinx. The processor has 16-bit data paths, and its I/O interface was extended with a *port_busy* signal that allows an external I/O device to stall the processor. These two modifications greatly simplified controller firmware programming and improved the overall speed.

The JCU module has four interfaces (see Fig. 8.12):

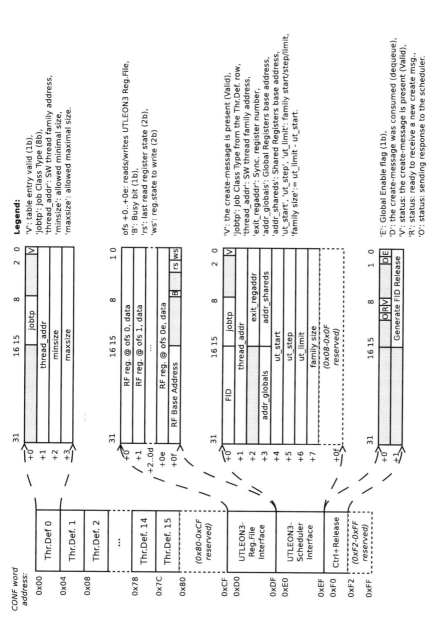

Fig. 8.19 The UTLEON3/HWT interface unit configuration address space

- The *CONF Slave* interface allows the main UTLEON3 processor to control the JCU during system initialization.
- The *Bank P* port connects to the local memory bank ('BRAM 3') that contains the firmware for the simple CPU. This is not needed for the JCU-FSM version.
- The *CX-Master* port allows the JCU to access the configuration registers of all the other modules in the subsystem.
- The *BX-Master* port allows the JCU to directly read/write the local memory banks. This functionality is not currently used in the demonstrator design.

8.7.6 Putting It All Together

To validate the design described a simple VADD program (Sect. 5.2) was used as a basis for the demonstration purposes.

The microthreaded assembler source code is displayed in Listing 2. The family of threads ut_vadd computes an integer vector addition, i.e. $Z_i := A_i + B_i$. As the same functionality is available in the *simple BCE core* (Sect. 8.7.3), we used it as one configuration of the hardware accelerator.

The program flowchart is shown in Fig. 8.20. The UTLEON3 processor starts in the legacy mode. After ensuring the JCU and BCE are stopped (step L1), the DMA Engine is used to download the appropriate JCU firmware to the local memory 'Bank P' (step L2). In the meantime the *Thread Mapping Table* (Sect. 8.7.4) is set up: one of its entries is filled up with the address of the software thread ut_vadd (step L3). Then the JCU is enabled (step L4), and the processor enters the microthreaded mode (step L5). All these configuration accesses are done by the UTLEON3 processor through the *Host Bridge*.

In the microthreaded mode the ut_vadd family is about to be created. First, its global registers %tg0–%tg2 are set up (step UT1); they contain pointers to the two input arrays (%tg0, %tg1) and the output array (%tg2). Then the family is allocated and set up (step UT2), and finally created (step UT3). At this time the processor scheduler sends a create query to the *UTLEON3 Processor Interface*. The interface device consults the thread mapping table and tries to pair the software family with one of the hardware configurations. The look-up process is based on the thread starting address that was specified in the setthread instruction. In our case the look-up is successful as the same address was written to the table in step L3. Were it not the case, an unsuccessful create query would result in the family of threads being executed in software as usual.

In the meantime the JCU is awaiting a create message from the *UTLEON3 Processor Interface* (step H1). When the create message is delivered, the JCU loads the family global registers and some other parameters (step H2). The DMA Engine is set up and engaged to download the two input arrays of the VADD program from the main memory to the local memory Banks A and B (BRAMs 0, 1) (step H3). Afterwards the selected BCE core is configured (step H4) and run (step H5). While

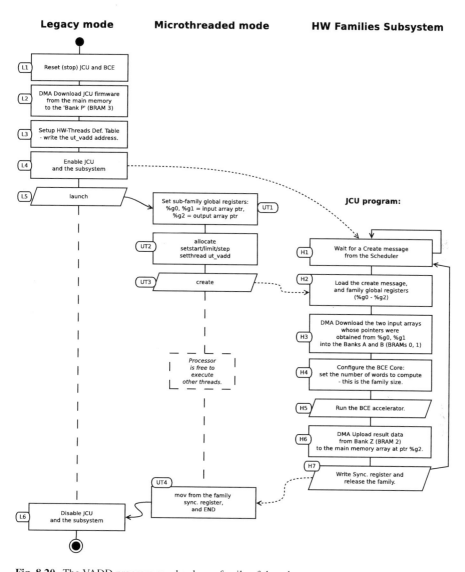

Fig. 8.20 The VADD program as a hardware family of threads

the BCE accelerator is computing, the JCU can theoretically serve other requests. Once the computation in the BCE has finished, the DMA Engine uploads the results from the local memory Bank Z (BRAM 2) to the main memory at address %tg2 (step H6). Finally, the JCU writes the family synchronization register (step H7), and signals the processor scheduler to release the family from the *Family Table*.

8.8   Implementation Characteristics

The crucial point in efficient use of hardware families of threads is the efficiency of accessing the Thread Mapping Table (TMT) during the family *create* process. This section evaluates how UTLEON3 can tolerate different TMT access latencies and its behaviour on a microthreaded implementation of a FIR filter shown in Listing 11.

The program was transformed by hand into the *mtsparc* assembler for execution in the microthreaded processor UTLEON3. Simultaneously an accelerated implementation in a hardwired BCE core was developed. The FIR BCE core was instantiated twice in the HWFAM subsystem to allow a concurrent execution of up to two families of threads in hardware. The JCU allocates a FIR BCE for a particular instance of a family of threads dynamically. In our implementation the JCU temporarily switches off the TMT unit while both the BCE cores are occupied. However, other job scheduling schemes can be implemented in the JCU firmware.

Listing 11 Program FIR in the microthreaded C language

```
   /* Microthreaded MAC (Multiply-Accumulate)   */
2  thread mac(shared int sum, int *x, int *y)
   {
4      index i;   /* local thread index, assigned
               according to the create() at line 14 */
6      sum = sum + x[i] * y[i];
   }
8
   /* Computes FIR at point z[k]. */
10 thread fir1(int *x, int *b, int *z, int t)
   {
12     index k;
       int fid, s = 0;
14     create (fid; start=0; step=1;
               limit=t-1; )
16        mac(s, &x[k], b);
       sync(fid);
18     z[k] = s;
   }
20
   /* Computes FIR for the whole vector z[] */
22 void fir(int *x, int *b, int *z,
           int t, int n)
24 {
       int fid;        /* fid = Family ID */
26     create (fid; start=0; step=1;
               limit=(n-t); )
28        fir1(x, b, z, t);
       sync(fid);
30 }
```

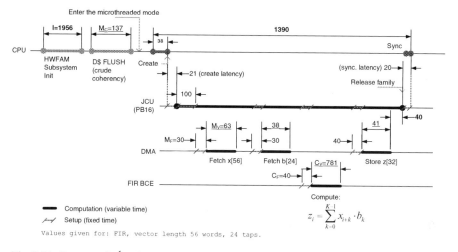

Fig. 8.21 Data transfers and computation in the hardware families of threads

In the experiments we chose the number of FIR taps to be constant ($t = 24$) as this is a typical value used in the embedded DSP domain for signal processing tasks.

The whole design (UTLEON3 processor, HWFAM subsystem, peripheral modules) can be synthesised in the *Xilinx XUP-V5 Board* that is also supported in the GRLIB package. In the experiments the main system memory had a latency of two cycles per word as this is the required configuration of the external on-board 1 MB SRAM.

Figure 8.21 shows a representative breakdown of a computation in the hardware accelerator.

8.8.1 Synthesis Results

The UTLEON3 with the HWFAM system was synthesized using *Synopsis Synplify D-2010.03* and *Xilinx ISE 12.3* in *Xilinx Virtex 5* xc5vlx110t. Table 8.3 lists the synthesis results. Apart from the FIR BCE core, a hardware implementation of the *Discrete Cosine Transform* (DCT) from *JPEG* was developed; however, its performance is not discussed here further. The number of the TMT rows r governs the number of classes of families of threads (i.e. distinct functions) that can be mapped to hardware accelerators at the same time. In the presented experimental setup with two equivalent FIR BCEs we required only one TMT row that maps the symbolic address &(**thread** fir1) to a FIR BCE.

In our simple implementation of the TMT unit the content-addressable memory performs lookups in $O(1)$ time. However, from Table 8.3 we see that in terms of hardware resources the implementation does not scale well when the number of

Table 8.3 Synthesis results in *Xilinx Virtex 5* technology (XC5VLX110T). r = Number of rows in the associative Thread Mapping Table

Component	BRAM	FFs	LUTs	DSP48E
UTLEON3	86	5,405	10,874	4
DMA + Bridge + Internal bus	15	495	549	0
Job Control Unit	4	126	409	0
FIR BCE	6	216	114	3
DCT BCE	2	542	451	4
Thread Mapping Table				
r = 16 rows	9	915	644	0
r = 8 rows	8	665	529	0
r = 4 rows	8	539	495	0
r = 2 rows	8	475	455	0

TMT rows r increases. Other implementation techniques can improve the resource scaling issue, but probably at the cost of increased lookup latency; this issue is dealt with in the following section.

8.8.2 TMT Latency and Execution Efficiency

We evaluate the impact of the TMT latency on the overall processor execution. The processor must cope with additional latencies as all the *create events* are routed through the TMT unit. The latency tolerance can be measured in terms of the CPU efficiency, or IPC (*Instructions Per Cycle*):

$$IPC = \frac{IC}{CC} \qquad (8.2)$$

where *IC* is the instruction count, and *CC* is the total running time in clock cycles.

We ran eight identical FIR computations in software to simulate higher workload and to average out transient effects of the dynamic thread scheduling. We limited the number of concurrently scheduled FIRs to at most three by the *blocksize* parameter so that we obtain more realistic (less parallel) workload.

8.8.2.1 Baseline Efficiency Degradation

Table 8.4 shows the IPC degradation when the TMT unit is switched on. All *create events*, including those not belonging to any hardware accelerated family, are routed through the TMT unit, but no families of threads are actually executed in the accelerators (otherwise the instruction count IC would no longer be constant). We see that in the worst case ($n = 26$) the IPC degradation is 2.5%-points, but typically it is less than 1%-point.

Table 8.4 CPU efficiency IPC degradation due to the added latency of the TMT unit; n is FIR input vector length; number of taps: $t = 24$

n	CPU IPC		IPC Degradation
	no TMT	*with TMT*	
26	0.659	0.634	−0.025
32	0.665	0.661	−0.004
40	0.676	0.673	−0.002
48	0.681	0.676	−0.005
64	0.680	0.673	−0.007
80	0.679	0.673	−0.006

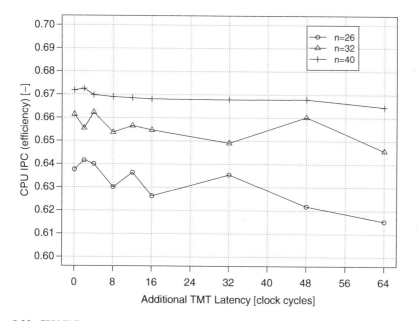

Fig. 8.22 CPU IPC degradation when TMT unit's latency increases

8.8.2.2   TMT Latency Scaling

So far we have assumed the TMT unit's latency to be only a few clock cycles. An implementation of the TMT unit with large number (tens to hundreds) of rows r would allow us to simultaneously map many classes of families of threads to hardware. One example is to map kernels from the *BLAS* package. However, in such an implementation the TMT latency will likely be in the order of $O(\log r)$, or even $O(r)$ if sequential search is used.

To simulate this situation we modified the TMT unit to impose an additional artificial non-pipelined latency during its processing of the *create events* coming from the processor; Fig. 8.22 plots the efficiency IPC with respect to the additional latency. In the worst case ($n = 26$) the IPC degrades from 0.640 to 0.615 when

Table 8.5 Extended measurement data for a 26-tap FIR filter with different vector lengths. Top – execution in UTLEON3 without HWFAM, middle – execution always in HWFAM, bottom – combined execution in UTLEON3 or HWFAM according to the vector length

n	Kernel cc	Coherency cc	CPU IC	DMA cc	FIR BCE cc	RF cc	Total cc	Speedup
26	849	0	403	0	0	0	849	1.0 (base)
40	3,344	0	2,111	0	0	0	3,344	1.0 (base)
48	4,752	0	3,087	0	0	0	4,752	1.0 (base)
128	19,028	0	12,847	0	0	0	19,028	1.0 (base)
26	2,620	584	37	226	84	9	3,204	0.26
40	2,957	729	37	352	420	9	3,686	0.91
48	3,143	799	37	424	612	9	3,942	1.21
128	5,702	1,536	37	1,151	2,532	9	7,238	2.63
26	905	0	403	0	0	0	905	0.94
40	3,429	0	2,111	0	0	0	3,429	0.98
48	3,143	799	37	424	612	9	3,942	1.21
128	5,702	1,536	37	1,151	2,532	9	7,238	2.63

TMT latency increases by 64 cycles. This shows that the microthreaded processors tolerates the latencies introduced by our coupling scheme and validates the approach (see also Table 8.5).

8.8.3 Family Size and Execution Efficiency

The FIR BCE computes one multiply-accumulate operation (MAC) per cycle while the software implementation requires at least four cycles for the same computation (2x LOAD, 1x MUL, 1x ADD). However, the main disadvantage of the hardware implementation is the need to transfer the I/O data to/from the main memory. In the experiments we considered the worst case scenario: at the beginning the data were live in the processor cache, they had to be flushed to the main memory, then downloaded into the BCE local buffer memory by the DMA engine, and at the end the results uploaded back to the main memory. Combined with other control overheads in the JCU the software implementation of a family can easily be faster.

Figure 8.23 shows a speedup of a *single* FIR computation with respect to the input vector length n. The number of the FIR taps is constant ($t = 24$). The computation requires $t \cdot (n - t + 1) = (nt - t^2 + t)$ MAC operations, and $(t + n + (n - t + 1)) = (2n + 1)$ words have to be transferred over the system bus and evicted/flushed from the processor D-Cache beforehand, which roughly doubles the actual number of clock cycles it takes. Indeed, the sample accelerated hardware FIR BCE reaches a speedup of 2x for $n = 96$ (i.e. 1,752 MACs); however for $n < 44$ (i.e. less than 504 MACs) the software implementation is faster due to the data transfers (the ∘ plot has the speedup much less than 1 in that region). The performance of the HWFAM coupling scheme can be improved by filtering families of threads in the TMT unit according to their size so that short FIR families will be always executed

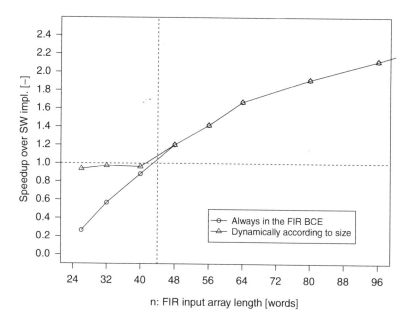

Fig. 8.23 Speedups (including data transfers) over all-software implementation for short input vectors n where accelerator overheads are most pronounced. Single FIR BCE, $t = 24$ taps, $24 \cdot (n - 23)$ MAC operations

in software. This is shown by the $\triangle$ plot which almost restores the performance in the troublesome region. The remaining speedup degradation is due to the added latency of the *create events* introduced in the TMT unit.

8.8.4   HWFAM Execution Profile

We will conclude the characterisation with a pipeline execution profile for a 25-tap FIR filter with loop unroll factor 4 (also see Sect. 7.1.2). The measured data are shown in Table 8.6 and Fig. 8.24. As in the previous performance tables the first column shows the execution core. The second column contains the overall number of clock cycles. The remaining columns contain cycle breakdown. Four processor pipeline states and three HWFAM states were considered:

- Working – processor pipeline is processing useful instruction
- Holdn(cache) – processor pipeline is idle due to external data dependencies
- Stall – processor pipeline is idle due to internal data dependencies
- UT ovhd – a processor pipeline is idle due to thread management overhead
- HW DMA – HWFAM data transfers between the external DDR and internal BRAMs

Table 8.6 Pipeline execution profile for a 26-tap FIR filter, loop unroll factor 4 – LEON3 vs. UTLEON3 vs. HWFAM

Core	Total [1]	Working [1]	Holdn (cache) [1]	Stall [1]	UT ovhd [1]	HW DMA [1]	HW ovhd [1]	HW working [1]
LEON3	7,405	4,105	196	3,104				
UTLEON3	4,353	3,890	10	95	358			
HWFAM	1,390					242	289	859

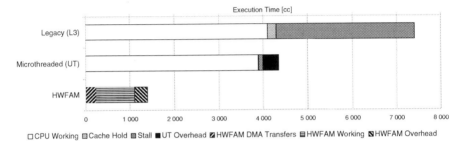

Fig. 8.24 Execution profile – HWFAM compared to LEON3 and UTLEON3

- HW ovhd – overhead due to setting up HWFAM control registers
- HW working – HWFAM performing the user computation

We can see that the HWFAM can execute the computation about 5x faster than LEON3 and about 3x faster than UTLEON3 including the data accesses to the external DDR memory.

8.9 Summary

The proposed microthreaded scheme for coupling reconfigurable accelerators to processors can be used with any processor that implements the microthreaded computing model. The benefit of the microthreaded coupling is that it handles concurrency in a uniform manner for both software and hardware execution.

In the performance evaluation we have intentionally focused only on specific features of the proposed design and evaluated the possible negative effects of increased latencies in the processor on its performance. The filtering of short families of threads in the TMT unit ensures that the custom hardware accelerators are not invoked if the expected overheads outweigh the speedup. On the other hand, it was shown that in the worst case the coupling scheme will degrade the processor efficiency by 2.5%-points in our case study, but typically less than 1%-point.

The performance data indicate external accelerators can be connected to the microthreaded pipeline efficiently enough to take over simple data-intensive tasks without degrading the integer pipeline performance.

Chapter 9
I/O and Interrupt Handling
in the Microthreaded Mode

An I/O (Input/Output) subsystem is provided by standard peripherals connected via the AMBA bus to the UTLEON3 processor. From the user perspective the peripherals are available in the non-cacheable address space. In the microthreaded mode a common peripheral access is provided in the same manner as in the legacy mode; the UTLEON3 processor does not require any extra modifications. The only difference is in interrupt (or trap) handling. The function of the peripherals in the legacy mode is described in the GRLIB IP Core User's Manual (see [5]). In the UTLEON3 processor interrupts are viewed as traps invoked by a peripheral. From the hardware point of view interrupts are handled by the multiprocessor interrupt controller (IRQMP [5]) in the legacy mode. The interrupt scheme is depicted in Fig. 9.1.

To handle interrupts in the microthreaded mode the UTLEON3 processor is extended with a new trap and interrupt controller (TIC). The block diagram of the UTLEON3 processor with TIC is shown in Fig. 9.2.

9.1 Trap Handling

The UTLEON3 processor supports traps due to external interrupts. The traps invoke special threads that service them, nevertheless the thread execution principles are the same as what has been described in the previous chapters. Trap handling requires us to introduce two new events in the UTLEON3 processor pipeline:

- Registration of a thread as a trap (or interrupt) handler, and
- Invocation of a thread when a trap (or interrupt) occurs.

These events are recognised and handled in a new trap and interrupt controller and the thread scheduler.

M. Daněk et al., *UTLEON3: Exploring Fine-Grain Multi-Threading in FPGAs*,
DOI 10.1007/978-1-4614-2410-9_9, © Springer Science+Business Media, LLC 2013

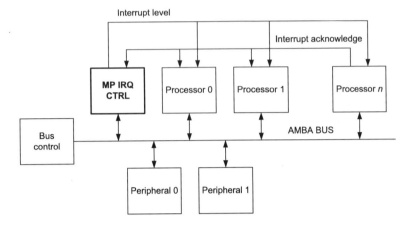

Fig. 9.1 UTLEON3 multiprocessor system with the multiprocessor interrupt controller [5]

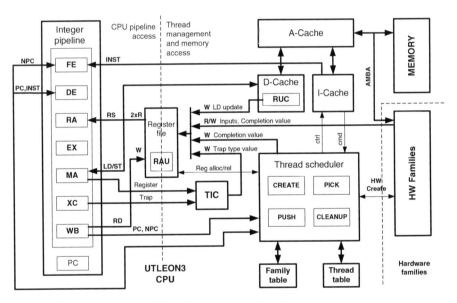

Fig. 9.2 Block diagram of the UTLEON3 processor with TIC

9.1.1 Thread Registration

A thread is registered as a trap handler through an execution of the `lda` instruction
on a specific address and address space. This `lda` instruction always results in a D-
Cache miss that marks the `lda` target register as pending as if a cache line fetch was
issued. The only difference from the usual cache miss is that the fetch request is not
generated. The instruction also updates the data structures in the TIC component.

When the thread reads the target register of the lda instruction, it gets *suspended*, and the target register is marked as *WAITING*; this enables the thread to be woken up later on.

9.1.2 Thread Invocation

A trap is triggered by the integer pipeline (IU3). The IU3 passes the trap request to the thread scheduler. The scheduler forces a thread switch, and forces the IU3 to execute nop operations (*force-nop*) in the pipeline to prevent threads from entering the processor pipeline. The IU3 passes the trap request to TIC as well. If there is a handler thread that is registered for the given trap type, the corresponding register is updated, which generates a high-priority thread wakeup request. The thread scheduler checks if the the handler code is present in the I-Cache, and issues a cache line fetch request if necessary. If the handler code is present in the I-Cache (or the fetch request has been completed), the scheduler bypasses the regular list of active threads and sets the integer pipeline to execute the handler thread.

9.1.3 Thread State Transitions

Threads used as trap handlers are managed in the same way as normal threads. The only difference is in the thread wakeup procedure mentioned above. A short execution time of a trap handler execution can be ensured only if:

- The part of the thread handler that is to be executed between a thread wakeup and subsequent thread registration fits in one I-Cache line,
- No long-latency operation is used in the trap handler,
- No swch instruction modifier is used in the trap handler.

If the conditions mentioned above are met, the thread handler state will change as shown in the left part of Fig. 9.3. A *running* thread becomes *suspended* at the time it registers itself for a given trap type and accesses the target register of the corresponding lda operation. When a trap occurs, the thread may become *waiting* for a short time until the I-Cache has fetched the required cacheline, then it becomes *running* immediately. The thread becomes *suspended* again at the time it re-registers itself for the next trap event.

If the conditions are not met, the thread handler state changes will be more complex as shown in Fig. 9.4. The trap handler run time will not be deterministic, and short execution time cannot be guaranteed.

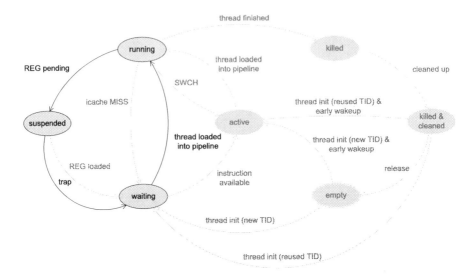

Fig. 9.3 Thread state transitions – short-latency trap handler

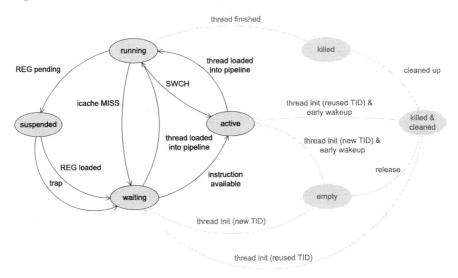

Fig. 9.4 Thread state transitions – general trap handler

9.2 Trap and Interrupt Controller

The trap and interrupt controller (TIC) provides two basic functions:

- Handler thread registration, and
- Handler thread wake up.

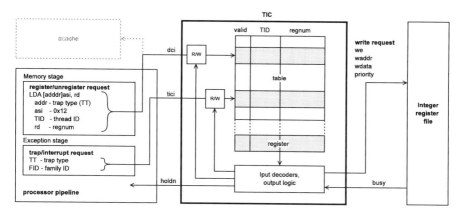

Fig. 9.5 Simplified block diagram of the trap and interrupt controller

TIC consists of a data structure and an auxiliary logic used to process input requests, to process data from the data structure, and to write data to registers in the register file through the shared asynchronous access port. The data structure is a table with lines consisting of three items:

- Line valid indicator,
- Thread ID (TID) of the handler thread, and
- Physical number of the corresponding register (regnum).

The table is indexed by the trap type (TT) register that identifies the current trap to be handled. The SPARC V8 specification defines up to 256 traps with the corresponding trap type values from 0x00 to 0xFF. Inside the range there is a block of values that are reserved as implementation dependent; the block starts with 0x60 and it ends with 0x7F. TIC supports registration of a universal handler that serves all unregistered traps. A particular trap type assigned to the universal handler as a placeholder is equal to 0x60. As the UTLEON3 processor uses trap types from 0x00 to 0x3F (6 bit index), TIC has a special register to store the TID of the universal handler. This register has the same structure as the table line.

As the registration and wake-up operations can access the trap table at the same time, and both of them need to prefetch the previous value before writing, the table needs up to four ports. Figure 9.5 shows a simplified block diagram of TIC. Assuming the use of a dual-port memory (BlockRAM entity in the FPGA) the TIC operations will take two clock cycles. In the case of successive operations of the same type in subsequent clock cycles the *holdn* signal becomes active and halts the processor pipeline until TIC is ready to accept new operations. The processor pipeline is also halted when registration and wake-up requests are generated in the same clock cycle because of the shared access to the integer register file.

9.3 Interrupt Registration

A thread is registered as an trap handler by executing the lda instruction that accesses a specific address in a specific address space:

```
lda [ADDR] ASI, %rd,
```

where

- ADDR is the address that corresponds to the trap type (TT), the address is shifted by 4 bits – the first 2-bit shift ensures memory address alignment, the second 2-bit shift is to improve readability of the trap types in the assembler code,
- ASI is the address space identifier (ASI) 0x12, and
- %rd is the destination register related to the trap handler thread that is used as a synchronizing register.

TIC detects registration transactions by monitoring the D-Cache input interface (dci) (see Fig. 9.6). The ci values are used to write data to the corresponding table line or special register; this depends on the current trap-type value. The registration takes two clock cycles. In the first clock cycle the required line from the table (given by the trap type) or the register (when the trap type is 0x60) are read to obtain the valid bit. In the second clock cycle new values are written to the table or the register, and a register file request is generated according to the value of the valid bit:

- 0 – a write request to the register file is not generated; this means that no trap handler is registered for the requested trap type. The table line or the register stay unchanged.
- 1 – a write request to the register file is generated only when the current and new TIDs are not equal. This means that the re-registration operation is required by another thread for the input trap type. The written data is marked as *KILLED* (0xFFFFFFFF), the state written is *FULL*. This write request is not prioritized. The valid flag becomes reset in this case.

Figure 9.7 shows the waveform on the register file interface.

Fig. 9.6 Transaction on *dci* with a TIC register

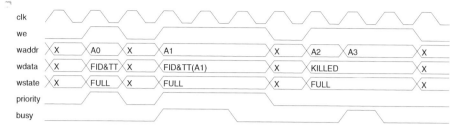

Fig. 9.7 Register file request interface

9.4 Interrupt De-registration

A trap handler de-registration is initiated on execution of the lda instruction when the destination register is %r0. The input trap type determines whether the corresponding line valid flag or register valid flag are reset. This operation initiates a write request to the register file with the *KILLED* value as well as the de-registration operation.

9.5 Interrupt Request

An interrupt request is processed in the exception stage of the processor pipeline. It consists of the TT number related to the interrupt request and the family identifier (FID) of the interrupted family. The input interrupt request interface is shown in Fig. 9.8. The operation in TIC takes two clock cycles. It consists of getting the information about the registered handler (TT matches the line from the trap table or the special register), updating the table or the special register, and updating the corresponding register in the integer register file. In the first clock cycle the corresponding line from the table and the register is read. If the table line is invalid, the register will be used. The value of the valid bit determines the behaviour during the second clock cycle:

- 0 – the interrupt request is ignored; this means that neither the general, nor a specific interrupt handler is registered for the requested trap type. The table line and the register are left unchanged.
- 1 – the interrupt request is taken. The valid bit is reset, and a write request to the register file is generated; the data to be written consist of the input FID (Family ID) and the trap type, the new register state is *FULL*. This write request is accompanied with a high-priority wake up request to the scheduler. Figure 9.7 shows the register file interface waveform on interrupt request.

The TIC block diagram is shown in Fig. 9.9.

Fig. 9.8 Interrupt request
interface

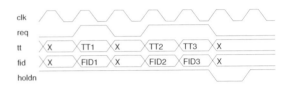

9.6   Interrupt Handler Example

An interrupt handler for a given trap type is formed by a family of threads with just
one thread. The interrupt handler family is created via the regular family creation.
The body of a handler thread within the family performs a sequence of operations
outlined below:

1. A thread is registered as an interrupt handler through the lda instruction
   that accesses a specific address in a specific address space (lda [ADDR]
   ASI, %rd). The address space reserved for TIC is 0x12. Individual addresses
   correspond to trap type identifiers shifted by 4 bits to the left. The destination
   register %rd acts as a synchronizing register.
2. The UTLEON3 trap/interrupt subsystem is enabled when the *ET* bit in the
   PSR register is set, and when the interrupt request of the interrupting device is
   unmasked in the multiprocessor interrupt controller.
3. The thread becomes *suspended* when the synchronizing register is read.
4. On an interrupt request the thread is woken up and the synchronizing register is
   read to determine if the thread is to be ended (the *KILLED* wakeup value) or the
   interrupt routine executed.
5. The interrupt routine is executed.
6. Go to Step 1.

An example of a timer interrupt handler is shown in Listings 12 and 13, The
interrupt handler starts with the thread creation (lines 2–10). The code can continue
with another thread creation. When all regular threads are finished, the thread
that executes the interrupt handler is also finished (lines 12–20). The interrupt
handler body starts at line 24 with the initialization part; the registers are initialized
with values that will be constant during the handler execution. The main part of
the thread is formed by the loop, its body contains the handler registration and
interrupt routine. The interrupt is registered through the lda instruction (line 43).
Sequentially, traps are enabled and interrupt requests from the timer are unmasked
in the multiprocessor interrupt controller. After that the handler thread becomes
suspended until an interrupt request occurs (line 46). When the thread is woken
up, the value of the synchronizing register is evaluated. If the value corresponds to
the *KILLED* flag (0xFFFFFFFF), the thread will be ended, otherwise the interrupt
routine will be executed.

The routine uses a buffer of pending interrupts located in the main memory with a
pointer to the top of the buffer. To make the interrupt processing as short as possible
the routine only increments the pointer and writes the data read from the peripheral

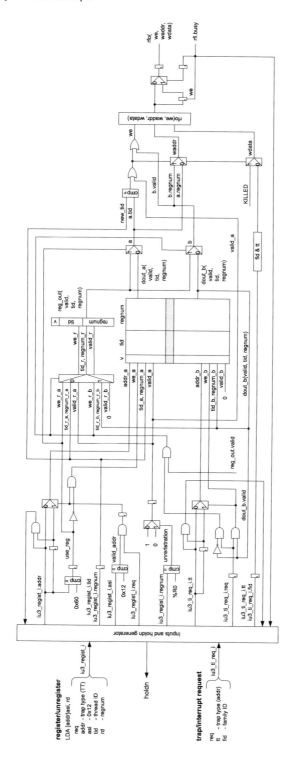

Fig. 9.9 Trap and interrupt controller block diagram

Listing 12 An `interrupt handler` example (mtsparc assembler listing)

```
   /* create timer interrupt handler family */
 2 allocate %tl23
   setstart %tl23 , 0
 4 setlimit %tl23 , 0
   setstep %tl23 , 1
 6 setblock %tl23 , BLOCKSIZE
   set fticl_start , %tl21
 8 setthread %tl23 , %tl21
   create %tl23 , %tl23
10
   ...
12
   /* unregister the handler family so that it can be finished */
14 set TT_TIMER, %tl1
   lda [%tl1]0x12 , %r0
16               /* unregister ~ release the fticl_start family */

18 mov %tl23 , %tl21
                 /* wait for family ftic_start to terminate */
20 nop

22 ...

24 /* timer interrupt handler family */
   .align CACHELINE
26     .registers 0 0 10  0 0 0           /* 0 GR,  0 SR, 10 LR */
   fticl_start:
28     set ET_MASK, %tl1
       rd %psr , %tl2
30     and %tl2 , %tl1 , %tl2
       wr %tl2 , %psr                             /* disable traps */
32     set TT_TIMER, %tl3      /* TT − interrupt level 8 − timer */
       set KILLED, %tl5
34     set INTCTRL_MASK_BASE, %tl1
       set TIMER_MASK_VAL, %tl7
36     set TIMER_UNMASK_VAL, %tl8
       set IRQ_BUFF_PTR , %tl9
38               /* Pointer to pending interrupt buffer */
```

to the buffer. The routine ends with a jump to the start of the inner loop to make a new registration in the TIC component to keep the handler alive. The data in the buffer are processed by another thread.

It should be noted that traps are disabled automatically when an interrupt request occurs in the processor pipeline. They must be enabled in each loop iteration.

Listing 13 An `interrupt handler` example (mtsparc assembler listing) cont'd

```
     /* ============================================================ */
40   .align  CACHELINE
     fticl_cont:                                     /* loop */
42       rd %psr , %tl2
         or %tl2 , 0x20 ,%tl2
44       lda [%tl3]0x12, %tl4                    /* register */
         wr %tl2 , %psr                          /* enable traps */
46       st %tl8 , [%tl1]                        /* unmask interrupt */
         mov %tl4 , %tl4    /* make thread PENDING until trap occurs */
48       cmp %tl4 , %tl5    /* is it 'KILLED' wakeup or TRAP wakeup? */
         be fticl_killed
50       ld [%tl9] , %tl6       /* delay slot - get current pinter */

52   fticl_handler:                            /* interrupt routine */
         inc %tl6                              /* calculate new pionter */
54       stb %tl3 , [%tl6]                     /* store TT to buffer */
         st %tl6 , [%tl9]                      /* store new pointer */
56       ba fticl_cont                         /* keep the thread alive */
         nop                                   /* delay slot */
58
     fticl_killed:                        /* end the thread -> end family */
60       st %tl7 , [%tl1]  ; END !          /* mask interrupt */
```

9.7 UTLEON3 Interrupt Subsystem Characteristics

The usage of the interrupt subsystem is limited by hardware constraints of the current UTLEON3 implementation. There are two possible scenarios of the UTLEON3 processor usage to be considered. First, the UTLEON3 processor works with data from the peripheral which interrupts the processor when the data are ready. The question is how often the processor can be interrupted without losing some of the interrupt requests. Second, when a computation has to be finished in certain time, we are interested in knowing how the computation runtime is influenced by the incoming interrupt requests.

9.7.1 Minimal Time Between Two Interrupt Requests

The most important characteristics of an interrupt service routine is the time it takes to process an interrupt request, in other words the minimum allowable amount of time between two consecutive interrupt requests in a row such that no interrupt request will be lost (t_{IRQ_min}). There are three reasons why an interrupt request can be lost:

Listing 14 A computing thread example (mtsparc assembler listing)

```
   . align  CACHELINE
2      . registers  2 0 2  0 0 0                    /*  2 GR,   0 SR,  2 LR  */
   f 1 _start :
4      sll %tl0 , 2, %tl1

6      /* number of repeated instruction */
       . rept COMP_INSTR_CNT
8          inc %tl1
           dec %tl1
10     . endr

12     st %tl0 , [%tg1+%tl1 ]   ; END !
```

- The time between two consecutive interrupt requests is less than t_{IRQ_min}.
- The interrupt handler code is not loaded in the I-Cache when an interrupt request occurs.
- A handler is not registered for the request.

To get t_{IRQ_min} it is assumed that the interrupt handler is locked in the I-Cache, and the data buffer and the pointer to the top of this buffer are always available in the D-Cache. The number of instructions in the loop inside the handler (see Listing 13) has to fit in one cache line. An example program we will use to evaluate the TIC properties consists of one family of threads and one interrupt handler for the timer; the computation is interrupted by the timer. Listing 14 shows the thread body; it just sets an output vector to 0, 1, 2, 3, etc. The COMP_INSTR_CNT compile-time constant located on line 7 gives the number of instruction repetitions in its body to the compiler; the body consist of *inc* and *dec* instructions. The more the instructions, the longer the thread will execute. The additional instructions will not affect the results. In the measurements the constant was set to 20. The interrupt handler is the same as was presented in Sect. 9.6 (see Listings 12 and 13).

This program was compiled and then executed for several different settings of the timer. The time between two interrupt requests was decreased from 88 to 42 clock cycles. Figure 9.10 shows the experimental results. If the time is shorter than 58 clock cycles, the system is not able to response to all interrupt requests.

9.7.2   Impact of Interrupt Handling on the Program Runtime

The interrupt handling negatively influences the computation time; it makes the runtime longer. Equation 9.1 shows a runtime dependency on the time spent in the interrupt handler:

$$t_{total} = tc + \sum_{i=0}^{N} th_i \qquad (9.1)$$

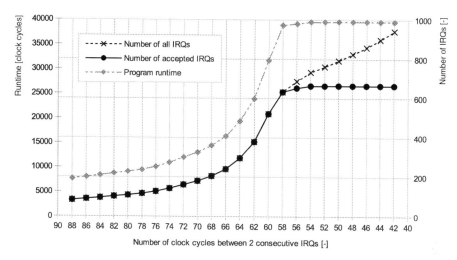

Fig. 9.10 The minimal time between two consecutive interrupt requests t_{IRQ_min} =58 clock cycles

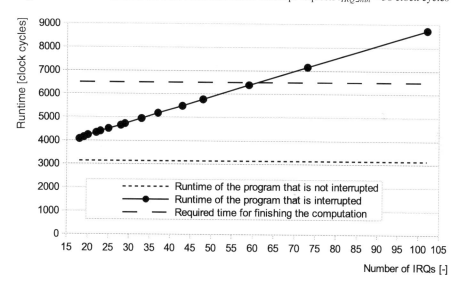

Fig. 9.11 The impact of interrupt handling on the program runtime

where tc is the time taken to execute the program without interrupts, and th_i is the time required for handling one interrupt request. In the tests we used exactly the same program as described in Sect. 9.7.1. The only difference is that the time between two interrupt requests was decreased from 220 to 80 clock cycles. We set the time required to finish the computation to 6,500 clock cycles. The results are shown in Fig. 9.11. The runtime increases linearly with the increased interrupt count. To complete the computation in the required time limit, it must not be interrupted more than 60 times.

9.8 Summary

The proposed microthreaded scheme for handling interrupts can be used with any processor that implements the microthreaded computing model. The benefit of the interrupt handling in the microthreaded mode is that there is no need to switch the UTLEON3 processor from the microthreaded to the legacy mode.

There are certain limitations on the implementation of trap handlers that are given by the current implementation of UTLEON3. The first is that in the current UTLEON3 the main body of a trap handler has to fit in one I-Cache line. This can be removed by implementing a mechanism in the I-Cache that would allow to explicitly lock certain cachelines used by handlers. The second limitation, rather an implementation characteristics, is the minimal time between two consecutive interrupt requests that can be processed by UTLEON3. At present it is 58 clock cycles; more frequent events will be lost.

Appendix A
The *IU3* Pipeline

A.1 IU3 Pipeline Register

The Table A.1 shows the basic register structure that implements the iu3 pipeline. Different table columns denote different pipeline stages (FE, DE, RA, EX, MA, XC, WB), each row lists register names for accessing one specific information in a given pipeline stage if it exists.

M. Daněk et al., *UTLEON3: Exploring Fine-Grain Multi-Threading in FPGAs*,
DOI 10.1007/978-1-4614-2410-9, © Springer Science+Business Media, LLC 2013

Table A.1 LEON3 (IU3) – structure of the pipeline register

Fetch	Decode	Register access	Execute	Memory access	Exception	Write back	Comments
F	D	A	E	M	X	W	
pc	pc	ctrl.pc	ctrl.pc	ctrl.pc	ctrl.pc		program counter
branch							The previous instruction was JUMP, CALL, TRAP instruction
	inst	ctrl.inst	ctrl.inst	ctrl.inst	ctrl.inst		instruction
	cwp	cwp	cwp			s.cwp	current window pointer
	set						Current set of multi-set instruction cache
					set		Current set of multi-set data cache
	mexc				mexc		Memory exception
	cnt	ctrl.cnt	ctrl.cnt	ctrl.cnt	ctrl.cnt		step counter and halt for UMUL/UDIV/FPU/CPU
	pv	ctrl.pv	ctrl.pv	ctrl.pv	ctrl.pv		pipeline valid
	inull	ctrl.trap	ctrl.trap	ctrl.trap	ctrl.trap		invalid instruction – trap, error,...
	annul	ctrl.annul	ctrl.annul	ctrl.annul	ctrl.annul		the previous instruction (stage A.E) was RETT and not JMPL and was not annulled
	step	step					the procesor is in the step mode (debug unit)
		ctrl.rd	ctrl.rd	ctrl.rd	ctrl.rd		address of the destination register (reg, ASR,...)
		ctrl.tt	ctrl.tt	ctrl.tt	ctrl.tt		trap type
		ctrl.wreg	ctrl.wreg	ctrl.wreg	ctrl.wreg		the instruction will write a result to a register

			Description
ctrl.wicc	ctrl.wicc	ctrl.wicc	the instruction will affect the flags – N,Z,O,C (PSR.ICC)
ctrl.wy	ctrl.wy	ctrl.wy	the instruction will write a result to the Y register
ctrl.ld	ctrl.ld	ctrl.ld	load
ctrl.rett	ctrl.rett	ctrl.rett	the instruction RETT (return from trap) is in the pipeline
rs1			source register 1
rfa1			register file address (register1)
rfa2			register file address (register2)
rsel1			operand 1 type (mux selector) – register, value, previous result, …
rsel2			operand 2 type (mux selector) – register, value, previous result, …
rfe1			register file enable (register 1)
rfe2			register file enable (register 2)
imm			immediate data
ldcheck1			load/icc interlock

(continued)

Table A.1 (continued)

| Fetch | Decode | Register access | Execute | Exception | Memory access | Write back | Comments |
F	D	A	E	X	M	W	
		ldcheck2					load/icc interlock
		ldchkra					load/icc interlock
		ldchkex					load/icc interlock
		su	su		su	s.s	supervisor/user mode (PSR.S)
		et	et			s.et	enable traps (PSR.ET)
		wovf					window overflow
		wunf					window underflow
		ticc					trap on integer condition code
		jmpl	jmpl				jump and link instruction
		mulstart					MUL start
		divstart					DIV start
			op1				operand 1
			op2				operand 2
			aluop				ALU operation
			alusel				ALU result select
			aluadd				ADD/SUB select
			alucin				ALU carry-in
			ldbp1				load bypass enable
			ldbp2				load bypass enable

Signal					Description
invop2					invert operand 2
shcnt					shift counter – bit count
sari					arithmetic shift – shift msb
shleft					shift left/right
ymsb					MULScc Y(msb)
rd					
icc	prev(m.icc)	icc	icc	s.icc	integer condition code
mulstep					MULScc Multiply Step (and modify icc)
mul					
mac		mac	mac		MAC instruction (UMAC or SMAC)
		result	result	result	result
		y		s.y	mul/div reg
		nalign			not aligned
		dci	dci		data cache input
		wcwp			write cwp
		irqen			enable interrupt request
		irqen2			
		divz			division by zero

(continued)

Table A.1 (continued)

Fetch	Decode	Register access	Execute	Memory access	Exception	Write back	Comments
F	D	A	E	M	X	W	
					annul_all		
					data		
					laddr		
					rstate		
					npc		next program counter
					intack		
					ipend		
					debug		
					nerror		
						wa	write register address
						wreg	enable write to register
						except	
						s.tt	trap type
						s.tba	trap base address
						s.wim	window invalid mask
						s.pil	processor interrupt level
						s.ec	enable CP
						s.ef	enable FPU
						s.ps	previous state of the supervisor flag
						s.asr18	
						s.svt	single vector trapping
						s.dwt	disable write error trap

Appendix B
Excerpts from the LEON3 Instruction Set

B.1 LEON3 (SPARC V8) Instruction Formats

Table B.1 Format 1 (op=1): CALL

31 30	29 0
op	disp30

Table B.2 Format 2 (op = 0): SETHI and branches

31 30	29	28 25	24 22	21 0
op	rd		op2	imm22
op	a	cond	op2	disp22

Table B.3 Format 3 (op = 2 or 3): Remaining instructions

31 30	29 25	24 19	18 14	13	12 5	4 0
op	rd	op3	rs1	0	asi	rs2
op	rd	op3	rs1	1	simm13	
op	rd	op3	rs1	opf		rs2

M. Daněk et al., *UTLEON3: Exploring Fine-Grain Multi-Threading in FPGAs*,
DOI 10.1007/978-1-4614-2410-9, © Springer Science+Business Media, LLC 2013

Table B.4 Table of instructions (op = 2)

op3[3:0]	op3[5:4] 0	1	2	3
9			RDPSR	RETT
A	UMUL	UMULcc	RDWIM	Ticc
B	SMUL	SMULcc	RDTBR	FLUSH
C	SUBX	SUBXcc		SAVE
D				RESTORE
E	UDIV	UDIVcc		UMAC
F	SDIV	SDIVcc		SMAC

Table B.5 Table of instructions (op = 3)

op3[3:0]	op3[5:4] 0	1	2	3
2	LDUH	LDUHA		
8				
9	LDSB	LDSBA		
A	LDSH	LDSHA		
B				
C				
D	LDSTUB	LDSTUBA		
E				
F	SWAP	SWAPA		

B.2   *Format 3* Instructions

B.3   Description of Assembler Instruction in GNU AS

The operation code from an assembler instruction is structured as shown in the table of instruction in the *binutils/opcodes/sparc-opc.c* file. The structure of the table is defined in the *binutils/include/opcode/sparc.h* header file. Each entry contains the following fields:

- An assembler instruction
- Bits in the operation code which must be set
- Bits in the operation code which do not have to be set
- Valid arguments as a string in a specific format
- Instruction flags (delayed, conditional/unconditional branch, . . .)
- A bitmask of the allowed SPARC architectures

Appendix C
Relevant LEON3 Registers and Address Space Identifiers

C.1 LEON3 Registers

- *r* registers
- IU control/status registers

    - Processor State Register (PSR)
    - Window Invalid Mask (WIM)
    - Trap Base Register (TBR)
    - Multiply/Divide Register (Y)
    - Program Counters (PC, nPC)
    - Implementation-dependent Ancillary State Registers (ASRs)

C.2 LEON3 Processor State Register

C.2.1 Fields of the Processor State Register

```
    Bits       Field   Description
 b31-b28        impl   Identifier of implementation
 b27-b24         ver   implementation-dependent version
                           of implementation
 b23-b20         icc   Integer Condition Codes
                       b23 - N - negative
                       b22 - Z - zero
                       b21 - V - overflow
                       b20 - C - carry
 b19-b14                reserved
     b13          EC   Enable Coprocessor
     b12          EF   Enable FPU
```

M. Daněk et al., *UTLEON3: Exploring Fine-Grain Multi-Threading in FPGAs*,
DOI 10.1007/978-1-4614-2410-9, © Springer Science+Business Media, LLC 2013

```
        b11-b8       PIL    Processor Interrupt Level
          b7         S      Supervisor/user mode
          b6         PS     Previous Supervisor (S bit at the
                                                time of
                                                the most
                                             recent trap)
          b5         ET     Enable Traps
        b4-b0        CWP    Current Window Pointer
```

C.2.2   Identifiers of Known SPARC V8 Implementations

```
  PSRimpl   PSRver    Company - Part Number           Label Used
     0         0      Fujitsu - MB86900/1A            Fujitsu0
                        & LSIL - L64801
     1        0,1     Cypress - CY7C601               Cypress0
                        & LSIL - L64811
     1         3      Cypress - CY7C611               Cypress1
     2         0      BIT - B5010                     BIT0
     5         0      Matsushita - MN10501            Matsushita0

     F         3      Gaisler - LEON                  LEON3
```

C.3   LEON3 Ancillary State Registers

C.3.1   List of Ancillary State Registers

```
    #ASR     Description
    1-15     Reserved
      16     SEU protection of the IU/FPU register file (LEON3FT)
      17     Processor Configuration Register
      18     UMAC,SMAC least 32bits of result
      19     Power-down register
      20
      21
      22
      23
      24     HW breakpoint 0, address register
      25     HW breakpoint 0, mask register
      26     HW breakpoint 1, address register
      27     HW breakpoint 1, mask register
      28     HW breakpoint 2, address register
      29     HW breakpoint 2, mask register
      30     HW breakpoint 3, address register
      31     HW breakpoint 3, mask register
```

C.3.2 Processor Configuration Register: ASR17

```
      Bits    Field    Description
   b31-b28    INDEX    Processor index.
   b27-b18             Reserved.
       b17    CS       Clock switching enabled. Switching
                          between AHB/CPU freq. is available.
   b16-b15    CF       CPU clock frequency. CPU core runs
                          at (CF+1) times AHB frequency.
       b14    DWT      Disable write error trap.
                          tt=0x2b will be ignored.
       b13    SVT      Single vector trapping enable.
       b12    LD       Load delay - the pipeline uses
                          2-cycle LD otherwise 1-cycle LD.
   b11-b10    FPU      Type of FPU. noFPU/GRFPU/MeikoFPU/
                          FRFPULite
        b9    M        Optional MAC instruction is available.
        b8    V8       SPARC V8 multiply,divide instructions
                          are available.
     b7-b5    NWP      Number of implemented watchpoints.
     b4-b0    NWIN     Number of implemented windows (NWIN+1).
```

C.4   LEON3 Address Space Identifiers

```
      ASI    *    Address Space
     0x01    c    Forced cache miss
     0x02    c    System control registers (cache control register)
     0x08    s    User Instruction
     0x09    s    Supervisor Instruction
     0x0A    s    User Data
     0x0B    s    Supervisor Data
     0x0C    c    Instruction cache tags
     0x0D    c    Instruction cache data
     0x0E    c    Data cache tags
     0x0F    c    Data cache data
     0x10    c    Flush instruction cache
     0x10    m    MMU flush page
     0x11    c    Flush data cache
     0x13    m    MMU flush context
     0x14    m    MMU diagnostic D-Cache context access
     0x15    m    MMU diagnostic I-Cache context access
     0x19    m    MMU registers
     0x1C    m    MMU bypass
     0x1D    m    MMU diagnostic access

   * s-SPARC V8 specification, c-LEON3 cache, m - LEON3 MMU
```

Appendix D
Scheduler Example

The example shows how the thread scheduler works with family and thread table entries.

D.1 Linked Lists Example: Initial Conditions

- Family 0
  - The thread that creates Family 1 and waits until it is finished
- Family 1
  - The threads computes
  - Parameters
    - Family size: 8
    - Indexes: 10 .. 17
    - Block size: 6

M. Daněk et al., *UTLEON3: Exploring Fine-Grain Multi-Threading in FPGAs*,
DOI 10.1007/978-1-4614-2410-9, © Springer Science+Business Media, LLC 2013

D.2   Linked Lists Example: 1/8

It is assumed that the Thread 0 in the Family 0 is running. Thread 0 is setting parameters that are required for Family 1 creation.

- Family 0
  - TID 0 running
- Family 1
  - Empty

Globals pointers

E-HEAD	1
E-TAIL	6
A-HEAD	/
A-TAIL	/

Family Table

FID	TAIL
0	0
1	/

ICache

CL	HEAD	TAIL
0	/	/
1	/	/

Thread Table

TID	INDEX	FPREV	FNEXT	SNEXT
0	0	/	/	/
1	/	/	/	2
2	/	/	/	3
3	/	/	/	4
4	/	/	/	5
5	/	/	/	6
6	/	/	/	/
7	-	-	-	-

Fig. D.1 Transitional digram of the thread states

D.3   Linked Lists Example: 2/8

Thread 0 initiated creation of Family 1 and is suspended, threads from Family 1 have been just created, thread 1 is running.

- Family 0
    - TID 0 suspended
- Family 1
    - TID 1 running
    - TID 2 active
    - TID 3 active
    - TID 4 active
    - TID 5 active
    - TID 6 active

Globals pointers

E-HEAD	/
E-TAIL	/
A-HEAD	2
A-TAIL	6

Family Table

FID	TAIL
0	0
1	6

ICache

CL	HEAD	TAIL
0	/	/
1	/	/

Thread Table

TID	INDEX	FPREV	FNEXT	SNEXT
0	0	/	/	/
1	10	/	2	/
2	11	1	3	3
3	12	2	4	4
4	13	3	5	5
5	14	4	6	6
6	15	5	/	/
7	-	-	-	-

D.4   Linked Lists Example: 3/8

Thread 1 and Thread 2 are waiting for cacheline 0, Thread 3 is running.

- Family 0

    - TID 0 suspended

- Family 1

    - TID 1 waiting
    - TID 2 waiting
    - TID 3 running
    - TID 4 active
    - TID 5 active
    - TID 6 active

Globals pointers

E-HEAD	/
E-TAIL	/
A-HEAD	4
A-TAIL	6

Family Table

FID	TAIL
0	0
1	6

ICache

CL	HEAD	TAIL
0	1	2
1	/	/

Thread Table

TID	INDEX	FPREV	FNEXT	SNEXT
0	0	/	/	/
1	10	/	2	2
2	11	1	3	/
3	12	2	4	/
4	13	3	5	5
5	14	4	6	6
6	15	5	/	/
7	-	-	-	-

D.5   Linked Lists Example: 4/8

A context switch occurred in the running Thread 3. Thread 3 state has been changed to active, Thread 4 is running.

- Family 0
  - TID 0 suspended
- Family 1
  - TID 1 waiting
  - TID 2 waiting
  - TID 3 active
  - TID 4 running
  - TID 5 active
  - TID 6 active

Globals pointers

E-HEAD	/
E-TAIL	/
A-HEAD	5
A-TAIL	3

Family Table

FID	TAIL
0	0
1	6

ICache

CL	HEAD	TAIL
0	1	2
1	/	/

Thread Table

TID	INDEX	FPREV	FNEXT	SNEXT
0	0	/	/	/
1	10	/	2	2
2	11	1	3	/
3	12	2	4	/
4	13	3	5	/
5	14	4	6	6
6	15	5	/	3
7	-	-	-	-

D.6   Linked Lists Example: 5/8

Cacheline 0 has been loaded, Thread 1 and Thread 2 have been woken up.

- Family 0
    - TID 0 suspended
- Family 1
    - TID 1 active
    - TID 2 active
    - TID 3 active
    - TID 4 running
    - TID 5 active
    - TID 6 active

Globals pointers

E-HEAD	/
E-TAIL	/
A-HEAD	5
A-TAIL	②

Family Table

FID	TAIL
0	0
1	6

ICache

CL	HEAD	TAIL
0	⃝／	⃝／
1	/	/

Thread Table

TID	INDEX	FPREV	FNEXT	SNEXT
0	0	/	/	/
1	10	/	2	2
2	11	1	3	/
3	12	2	4	①
4	13	3	5	/
5	14	4	6	6
6	15	5	/	3
7	-	-	-	-

D.7   Linked Lists Example: 6/8

Threads with index 10 (Thread 1) and 11 (Thread 2) have been killed, Thread 1 has been cleaned up.

- Family 0
    - TID 0 suspended
- Family 1
    - TID 1 cleaned up
    - TID 2 killed
    - TID 3 active
    - TID 4 running
    - TID 5 active
    - TID 6 active

Globals pointers

E-HEAD	/
E-TAIL	/
A-HEAD	5
A-TAIL	3

Family Table

FID	TAIL
0	0
1	6

ICache

CL	HEAD	TAIL
0	/	/
1	/	/

Thread Table

TID	INDEX	FPREV	FNEXT	SNEXT
0	0	/	/	/
1	/	/	/	/
2	11	1	3	/
3	12	2	4	/
4	13	3	5	/
5	14	4	6	6
6	15	5	/	3
7	-	-	-	-

D.8   Linked Lists Example: 7/8

Thread 1 has been reused for thread with index 16, it changed the thread state to
active.

- Family 0

    - TID 0 suspended

- Family 1

    - TID 1 active
    - TID 2 killed
    - TID 3 active
    - TID 4 running
    - TID 5 active
    - TID 6 active

Globals pointers

E-HEAD	/
E-TAIL	/
A-HEAD	5
A-TAIL	(1)

Family Table

FID	TAIL
0	0
1	(1)

ICache

CL	HEAD	TAIL
0	/	/
1	/	/

Thread Table

TID	INDEX	FPREV	FNEXT	SNEXT
0	0	/	/	/
1	(16)	(6)	(/)	(/)
2	11	1	3	/
3	12	2	4	(1)
4	13	3	5	/
5	14	4	6	6
6	15	5	(1)	3
7	-	-	-	-

D.9   Linked Lists Example: 8/8

Thread 2 has been reused for thread with index 17, Thread 3 killed & cleaned up, Thread 4 killed.

- Family 0
  - TID 0 suspended
- Family 1
  - TID 1 active
  - TID 2 active
  - TID 3 cleaned up
  - TID 4 killed
  - TID 5 running
  - TID 6 active

Globals pointers

E-HEAD	3
E-TAIL	3
A-HEAD	6
A-TAIL	2

Family Table

FID	TAIL
0	0
1	2

ICache

CL	HEAD	TAIL
0	/	/
1	/	/

Thread Table

TID	INDEX	FPREV	FNEXT	SNEXT
0	0	/	/	/
1	16	6	2	2
2	17	1	/	/
3	/	/	/	/
4	13	3	5	/
5	14	4	6	/
6	15	5	1	3
7	-	-	-	-

Appendix E
Used Resources

E.1 Implementation Results

This section presents implementation results of the UTLEON3 processor from various aspects.

The first set of experiments provides synthesis results for the FPGA technology. First, the UTLEON3 processor is compared to the original LEON3 processor in four different configurations. Second, a breakdown of resource requirements per a particular component is provided. Finally, a breakdown of resource requirements for the hardware accelerator for families of threads is evaluated.

The second experiment evaluates the implementation of both processors in an FPGA in terms of their maximal operating frequency.

The third set of experiments analyzes implementation results for the ASIC technology. A comparison of overall resource requirements and circuit delay is provided.

In all the experiments four processor configurations were considered (see Table E.1). The configurations differ in six parameters – the I-Cache set size, D-Cache set size, D-Cache associativity level, Register File size (RF), Family Thread Table size (FTT) and Thread Table Size (TT). Note that the latter three parameters are valid for the UTLEON3 processor only. As a result, configuration 0 equals to configuration 2 and configuration 1 equals to configuration 3 for the LEON3 processor. A register file with one write port was considered in all cases.

M. Daněk et al., *UTLEON3: Exploring Fine-Grain Multi-Threading in FPGAs*,
DOI 10.1007/978-1-4614-2410-9, © Springer Science+Business Media, LLC 2013

Table E.1 Design
configurations for synthesis

CFG	Description
0	ICache 1 kB, DCache 1 kB, asociativity 1 RF 256, FTT 8, TT 64
1	ICache 1 kB, DCache 1 kB, asociativity 4 RF 256, FTT 8, TT 64
2	ICache 1 kB, DCache 1 kB, asociativity 1 RF 1024, FTT 32, TT 256
3	ICache 1 kB, DCache 1 kB, asociativity 4 RF 1024, FTT 32, TT 256

E.1.1   FPGA Synthesis

This section presents implementation results for the FPGA technology. The XST release 12.3, xst M.70d (lin64) was used for synthesis; the Virtex 5 LXT (xc5vlx110t-1-ff1136) FPGA was selected as the target.

E.1.1.1   Comparison of the LEON3 and UTLEON3 Processors

This subsection provides overall resource consumption for the LEON3 and UT-LEON3 processor cores. Four various configurations were considered as shown in Table E.1.

The synthesis results are shown in Table E.2. The first column contains configuration indexes (see Table E.1); the next three columns show resource requirements in terms of slice flip-flops, 6-input LUTs (LUT6) and dedicated 36 kbit RAM blocks (RAMB36) respectively. The upper part contains results for the original LEON3 processor. As this core does not contain microthreaded modules, configuration 0 is equal to configuration 2, and configuration 1 to configuration 3. The lower part contains results for the UTLEON3 core. The resources used are also evaluated as the ratio to the resource requirements of the LEON3 core of the same configuration.

Table E.2 FPGA synthesis – resource requirements summary. Configurations are described in Table E.1. Resource requirements are provided in terms of slice flip-flops, 6-input LUTs and dedicated 36 kbit RAM blocks

LEON3s						
CFG	Slice Flip Flops		LUT6		RAMB36	
0, 2	1,324		4,804		7	
1, 3	1,619		5,434		11	
UTLEON3s						
CFG	Slice Flip Flops		LUT6		RAMB36	
0	4,874	3.7x	12,413	2.6x	29	4.1x
1	5,243	3.2x	14,428	2.7x	36	3.3x
2	5,175	3.9x	12,937	2.7x	30	4.3x
3	5,544	3.4x	14,691	2.7x	37	3.4x

The UTLEON3 core is 3.6 times larger in terms of flip-flops, 2.7 times in terms of LUTs and 3.8 times in terms of RAM blocks on average.

E.1.1.2   Distribution of Resources

This section shows the distribution of resources between processor subblocks. The distribution for the original LEON3 processor is shown followed by the distribution for the UTLEON3 processor. Configuration 0 shown in Table E.1 was considered in both cases.

The functional blocks of the original LEON3s core included in this experiment are cache (CACHE), integer pipeline (IU3) and register file (RF). The resource distribution is shown in Table E.3. Note that the overall resource requirements for the whole processor (column LEON3) may differ from the sum of partial resource requirements of these subblocks. The first reason is that just three most important components are shown in the breakdown, but not all the components that the UTLEON3 core is composed of. The second reason is that a separate synthesis of the components cannot achieve some inter-component optimizations.

Table E.3 FPGA synthesis – distribution of FPGA resources between the LEON3 modules

Resource type	LEON3s	CACHE	IU3	RF
Slice Flip Flops	1,324	246	969	0
Total 6 input LUTs	4,804	1,651	2,805	8
Used as logic	4,765	1,651	2,766	8
Used as shift registers	39	0	39	0
Used as RAMs	0	0	0	0
BRAMs 36 kb	7	5	0	1

In the case of the original LEON3 processor the integer pipeline consumes most resources – about 58 % of overall requirements in terms of 6-input LUTs, and 73 % in terms of flip-flops. The integer pipeline consumes almost 1.7 times more than the cache subsystem in terms of 6-input LUTs and 1.13 more in terms on flip-flops. Resource requirements of the register file were negligible.

Resource distribution for the UTLEON3 processor is shown in Table E.4. The blocks of the microthreaded UTLEON3s core are the cache (UTCACHE), the integer pipeline (UTIU3), the register file (UTRF) and three new blocks: the family thread table (FTT), thread table (TT) and thread scheduler (SCHED).

In contrast to the original LEON3 core, the highest resource requirements came from the cache subsystem in the UTLEON3 core – about 50 % of all resources in terms of 6-intput LUTs, 44 % in terms of flip-flops and 45 % in terms of block RAMs.

Table E.4 FPGA Synthesis – resource requirements of the UTLEON3 blocks.

Resource type	UTLEON3s	UTCACHE	UTIU3	UTRF	FTT	TT	SCHED
Slice Flip Flops	4,874	2,123	1,478	292	31	0	850
Total 6 input LUTs	12,413	6,204	3,809	604	58	8	1,880
Used as logic	12,084	6,176	3,694	554	38	8	1,778
Used as shift registers	137	0	115	0	0	0	0
Used as RAMs	192	28	0	50	20	0	94
BRAMs 36 kb	29	13	0	1	7	7	1

E.1.1.3   Hardware Families of Threads

This subsection shows the FPGA synthesis results for the hardware accelerator for families of threads. The setting of the synthesis tool was the same as in the previous cases.

The hardware accelerator for families of threads was configured so as to support concurrent execution of 2x FIR tasks, 1x DCT task and 1x simple integer task.

The synthesis results are listed in Table E.5. The table shows resource requirements for the top-level entity (HWT03) that implements the hardware accelerator for families of threads and for its subblocks. These subblocks are the interface to the UTLEON3 processor (UT_IF), DMA Engine with internal bus infrastructure (COMM), Job Control Unit (JCU), FIR worker (FIR_BCE) and DCT worker (DCT_BCE),

Table E.5 Distribution of FPGA resources between the HWFAM modules

Resource type	HWT03	UT_IF	COMM	JCU	FIR_BCE	DCT_BCE
Slice Flip Flops	2,606	637	488	256	307	315
Total 6 input LUTs	5,031	884	519	672	457	593
Used as logic	4,225	568	487	468	393	539
Used as shift registers	12	0	0	0	0	12
Used as RAMs	794	316	32	204	64	42
BRAMs 36 kb	18	0	7	1	3	1

The interface to the UTLEON3 core (UT_IF) has the highest resource requirements in terms of flip-flops and 6-input LUTs.

E.1.2   FPGA Implementation

Both the processor cores were implemented in an FPGA on the Xilinx XUP-V5 board [20] in order to verify correct function of the cores and evaluate the maximal clock frequency. The maximal frequency was 110 MHz for the original LEON3 core, and 33.3 MHz for the new UTLEON3 core.

E.1.3   ASIC Synthesis

This section shows implementation results for the ASIC technology. The Synopsys
Design Compiler D-2010.03-SP3 was used for synthesis; the selected target technology was tcbn90ghptc (TSMC 90 nm general purpose high performance, typical
case timing).

E.1.3.1   Comparison of the LEON3 and UTLEON3 Processors

Four configurations were considered as in the case of the FPGA synthesis (see
Table E.1). A register file with one write port was used in all cases.

First, the area requirements in square millimeters are shown in Table E.6. The
UTLEON3 processor is 8.5 times bigger than the original LEON3 processor on
average.

Table E.6 ASIC synthesis –
resource requirements
summary. Configurations are
described in Table E.1.
Resource requirements are
shown in terms of square
millimeters

LEON3s		
CFG	Area	
0, 2	0.42	
1, 3	0.64	
UTLEON3s		
CFG	Area	
0	3.62	8.6x
1	4.78	7.5x
2	4.13	9.8x
3	5.29	8.3x

Second, the length of a critical path is shown in Table E.7. The UTLEON3
processor is 1.19 times slower than the original LEON3 core on average.

Table E.7 ASIC Synthesis –
critical path length [ns] for all
configurations described in
Table E.1

LEON3s		
CFG	Delay [ns]	
0, 2	2.06	
1, 3	2.09	
UTLEON3s		
CFG	Delay [ns]	
0	3.71	1.80x
1	3.69	1.77x
2	3.71	1.80x
3	3.72	1.78x

E.1.3.2   Distribution of Resources

The breakdown for all four configurations is in the Table E.8.

Table E.8 ASIC synthesis – area requirements of the UTLEON3 blocks in square millimeters

CFG	UTLEON3s	UTCACHE	UTIU3	UTRF	FTT	TT	SCHED
0	3.59	1.49	0.06	0.31	0.23	1.22	–
1	4.69	2.65	–	0.31	0.23	1.22	–
2	4.04	1.57	–	0.67	0.23	1.22	–
3	5.20	2.73	–	0.67	0.23	1.22	–

Appendix F
Tutorial

F.1 Overview

This tutorial shows an implementation of the microthreaded UTLEON3 processor on the *Xilinx XUPV5-LX110T Evaluation Board (XUP-V5)*. The described setup executes a microthreaded program that computes the first 16 Fibonacci numbers.

F.2 Scope

This document describes a microthreaded UTLEON3 template design customized for the *Xilinx XUPV5-LX110T Evaluation Board (XUP-V5)*, the board is shown in Fig. F.1. The template design is intended to help users become familiar with the UTLEON3 processor and the GRLIB IP library. It also includes a demonstration how to execute microthreaded programs.

F.3 Requirements

The following hardware and software components are required in order to use and implement the XUP-V5 UTLEON3 template design.

- UTGRLIB IP Library 1.0.17.
- PC with linux or Windows 2000/XP with Cygwin.
- XUP-V5 with the JTAG programming cable.
- RS232 serial null modem cable (Table F.1 shows serial terminal settings).
- Xilinx ISE 12.3 development software.
- Synopsys Synplify Pro 2011.03-SP2.

RS232 serial terminal settings are shown in Table F.1.

M. Daněk et al., *UTLEON3: Exploring Fine-Grain Multi-Threading in FPGAs*,
DOI 10.1007/978-1-4614-2410-9, © Springer Science+Business Media, LLC 2013

Table F.1 Serial terminal
settings

Setting	Value
Speed	115,200
Data bits	8
Stop bits	1
Parity	None
Flow control	None

Fig. F.1 Xilinx XUPV5-LX110T Evaluation Board (XUP-V5)

For the UTLEON3 software development the following tools are recommended:

- Modified *binutils* for the SPARC V8 processor.
- Modified *binutils* for the microthreaded SPARC processor.
- The *binrep* tool that changes the target platform identifier in an object file.
- GRMON v1.1.52.

UTLEON3 can be simulated with Mentor Graphics ModelSim 6.6e or GHDL.

F.4   Installation

The example design is included in the UTGRLIB IP library. The library is provided as a gzipped tar file. To extract it, execute

```
tar xzf ut-grlib-1.0.17.tar.gz
```

This will create a directory called ut-grlib-1.0.17 with all IP cores and template designs. On MS Windows hosts the extraction and all further steps should be made inside a Cygwin shell.

F.5   Template Design Overview

The template design is located in ut-grlib-1.0.17/designs/utleon-xup-v5. It consists of three key files:

- *config.vhd* – the VHDL package with design configuration parameters.
- *leon3mp.vhd* – the top level entity and instantiates all on-chip IP cores. It uses *config.vhd* to configure the instantiated IP cores.
- *testbench.vhd* – the testbench with external memory that emulates parts of the XUP-V5 board.

Each core in the template design is configurable with VHDL generics. The values of these generics are assigned from the constants declared in *config.vhd*, created e.g. with the *make xconfig* tool.

F.6   Simulation

The template design can be simulated in the testbench that emulates the prototype board. The testbench includes an external PROM that is pre-loaded with a test program. The test program will execute on the microthreaded UTLEON3 processor, and test various functions of the design. The test program will print diagnostics on the simulator console during the execution. Type the following commands to compile and simulate the template design and testbench in ModelSim:

```
make vsim
vsim -novopt testbench
```

A simulation is initiated in the transcript window, type

```
run -all
```

After that you should see a similar listing:

```
# Xilinx XUPV5 Evaluation Board
# GRLIB Version 1.1.0, build 4104
```

```
# Target technology: virtex5   ,  memory library: virtex5
# ahbctrl: AHB arbiter/multiplexer rev 1
# ahbctrl: Common I/O area disabled
# ahbctrl: AHB masters: 4, AHB slaves: 8
# ahbctrl: Configuration area at 0xfffff000, 4 kbyte
# ahbctrl: mst0: AppleCore                  AppleCore UTLEON3 Processor
# ahbctrl: mst1: Gaisler Research           AHB Debug UART
# ahbctrl: mst2: Gaisler Research           JTAG Debug Link
# ahbctrl: mst3: Gaisler Research           SVGA frame buffer
# ahbctrl: slv1: Gaisler Research           AHB/APB Bridge
# ahbctrl:       memory at 0x80000000, size 1 Mbyte
# ahbctrl: slv2: Gaisler Research           Leon3 Debug Support Unit
# ahbctrl:       memory at 0x90000000, size 256 Mbyte
# ahbctrl: slv3: European Space Agency   Leon2 Memory Controller
# ahbctrl:       memory at 0x20000000, size 512 Mbyte
# ahbctrl:       memory at 0x40000000, size 32 Mbyte, cacheable, prefetch
# ahbctrl: slv6: Gaisler Research           Generic AHB ROM
# ahbctrl:       memory at 0x00000000, size 1 Mbyte, cacheable, prefetch
# apbctrl: APB Bridge at 0x80000000 rev 1
# apbctrl: slv0: European Space Agency   Leon2 Memory Controller
# apbctrl:       I/O ports at 0x80000000, size 256 byte
# apbctrl: slv1: Gaisler Research           Generic UART
# apbctrl:       I/O ports at 0x80000100, size 256 byte
# apbctrl: slv2: Gaisler Research           Multi-processor Interrupt Ctrl.
# apbctrl:       I/O ports at 0x80000200, size 256 byte
# apbctrl: slv3: Gaisler Research           Modular Timer Unit
# apbctrl:       I/O ports at 0x80000300, size 256 byte
# apbctrl: slv4: Gaisler Research           PS2 interface
# apbctrl:       I/O ports at 0x80000400, size 256 byte
# apbctrl: slv5: Gaisler Research           PS2 interface
# apbctrl:       I/O ports at 0x80000500, size 256 byte
# apbctrl: slv6: Gaisler Research           SVGA frame buffer
# apbctrl:       I/O ports at 0x80000600, size 256 byte
# apbctrl: slv7: Gaisler Research           AHB Debug UART
# apbctrl:       I/O ports at 0x80000700, size 256 byte
# apbctrl: slv8: Unknown vendor             Unknown Device
# apbctrl:       I/O ports at 0x80000800, size 256 byte
# apbctrl: slv9: Gaisler Research           AMBA Wrapper for OC I2C-master
# apbctrl:       I/O ports at 0x80000900, size 256 byte
# apbctrl: slv10: Gaisler Research          General Purpose I/O port
# apbctrl:       I/O ports at 0x80000a00, size 256 byte
# apbctrl: slv12: Gaisler Research          AMBA Wrapper for OC I2C-master
# apbctrl:       I/O ports at 0x80000c00, size 256 byte
# utia apbperfcnt8: PERFCNT rev 1
# i2cmst12: AMBA Wrapper for OC I2C-master rev 2, irq 11
# grgpio10: 13-bit GPIO Unit rev 1
# i2cmst9: AMBA Wrapper for OC I2C-master rev 2, irq 14
# svgactrl6: SVGA controller rev 0, FIFO length: 384, FIFO part length: 128,
    FIFO address bits: 9, AHB access size: 32 bits
# apbps2_5: APB PS2 interface rev 2, irq 5
# apbps2_4: APB PS2 interface rev 2, irq 4
# gptimer3: GR Timer Unit rev 0, 8-bit scaler, 2 32-bit timers, irq 8
# irqmp: Multi-processor Interrupt Controller rev 3, #cpu 1, eirq 0
# apbuart1: Generic UART rev 1, fifo 8, irq 2
# ahbrom6: 32-bit AHB ROM Module,  164 words, 8 address bits
# ahbjtag AHB Debug JTAG rev 1
# ahbuart7: AHB Debug UART rev 0
# dsu3_2: LEON3 Debug support unit + AHB Trace Buffer, 8 kbytes
# utleon3_0: UTLEON3 processor rev 0
# utleon3_0: icache 1*4 kbyte, dcache 4*4 kbyte
```

```
# clkgen_virtex5: virtex-5 sdram/pci clock generator, version 1
# clkgen_virtex5: Frequency 100000 KHz, DCM divisor 2/5
```

F.7   Synthesis and Place&Route

The template design can be synthesized with Synplify Pro 2011.03-SP2. The synthesis can be executed in a batch mode or interactively. To use synplify in the batch mode, use the command:

```
make synplify
```

To use synplify interactively, use :

```
make scripts
synplify_pro leon3mp_synplify.prj
```

Next run place&route for the netlist generated with Synplify Pro:

```
make ise-synp
```

The final configuration bitstream is stored in the file *leon3mp.bit*.

F.8   Board Re-Programming

The XUP-V5 FPGA configuration PROM (xc5vlx110t-1-ff1136) can be programmed from the shell window with the following command:

```
impact -batch download.cmd
```

F.9   Software Development

The UTLEON3 processor supports both the legacy and microthreaded code execution. Two different toolchains are needed.

- **binutils** – standard binutils for the SPARC V8 architecture extended with the *launch* instruction.
- **mtbinutils** – modified binutils for the SPARC V8 architecture with the microthreaded extensions.

F.9.1   Legacy Code

The legacy code compilation requires only the standard binutils.

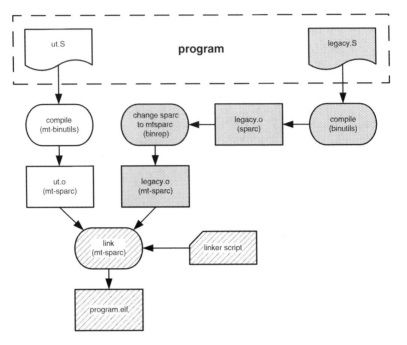

Fig. F.2 Microthreaded code compile flow

F.9.2   Microthreaded Code

As the microthreaded program code consists of both the legacy and microthreaded parts, the legacy and microthreaded parts have to be linked together. The necessary steps are as follows (see Fig. F.2):

1. Make an object file from the legacy assembler code with *binutils*
2. Make an object file from the microthreaded assembler code with *mtbinutils*.
3. Change the target platform code in the legacy object file from SPARC to microthreaded SPARC with the *binrep* tool.
4. Link the legacy and microthreaded object files into one.

A typical compile flow can be seen below.

```
sparc-elf-as legacy.s -o legacy.o
mtsparc-elf-as ut.s -o ut.o
./binrep legacy.o legacy_mod.o
mtsparc-elf-ld --script=linkprom-mt-allram legacy_mod.o ut.o -o
  out.elf --cref
```

F.10   Description of the Test Code

The sample program computes the first 16 Fibonacci numbers. The microthreaded part of the code is shown below.

```
/* CONSTANTS */
.equ   FIBNUM, 16
.equ   BLKSIZE, 4


.section ".text"
.global ut_main, __M0, __M1
.type ut_main, #function
.proc 04

.align 4*16
      .registers 0 0 31  0 0 0  /* Thread 0 cannot have any global, */
                                /*         local or shared registers */
ut_main:
__M0:
      set fibdata, %tl1   /* %tg1 */
      /* the first two fibonacci numbers: 1, 1 */
      mov 1, %tl8
      st %tl8, [%tl1]
      st %tl8, [%tl1+4]

      /* set first two fibonacci numbers for the first thread */
      mov 1, %tl8   /* %td0 */
      mov 1, %tl9   /* %td1 */

      allocate %tl20
      add %tl1, 8, %tl1
      /* start := 0 */
      setstart %tl20, 0
      /* limit := fibnum */
      set (FIBNUM*4-1), %tl22
      setlimit %tl20, %tl22
      /* step := 4 */
      setstep %tl20, 4
      /* blocksize */
      setblock %tl20, BLKSIZE
      /* thread := f1_start */
      set f1_start, %tl21
      setthread %tl20, %tl21
      /* and go */
      create %tl20, %tl20
      nop

      /* wait for family to terminate */
      mov %tl20, %tl21
      nop
__M1:
  nop ; END
.size ut_main, .-ut_main


.type f1_start, #function
.proc 04

.align 4*16
f1_addr1:
  .registers 8 2 2  0 0 0     ! GR, SR, LR
  /* %tg1 = address of fibdata */
  /* %tL0 = result word address [thread index] */
  /* %tD0,%tS0 = second-to-last fibonacci number [N-2] */
```

```
  /* %tD1,%tS1 = last fibonacci number [N-1] */
f1_start:
  add %td0, %td1, %tl1
  mov %td1, %ts0
  mov %tl1, %ts1
  st %tl1, [%tl0+%tg1]
  nop ; END !
.size f1_start, .-f1_start
```

F.11   GRMON

The whole system is controlled through the Aeroflex-Gaisler GRMON tool. Invoke the tool by typing:

```
grmon-eval -xilusb -ramws 2
```

You should see a similar listing in your terminal window:

```
GRMON LEON debug monitor v1.1.52 evaluation version

Copyright (C) 2004-2011 Aeroflex Gaisler - all rights reserved.
For latest updates, go to http://www.gaisler.com/
Comments or bug-reports to support@gaisler.com

This evaluation version will expire on 10/10/2012
Xilinx cable: Cable type/rev : 0x3
JTAG chain: xc5vlx110t xccace xc95144xl xcf32p xcf32p

Device ID: : 0x509
GRLIB build version: 4104

initialising ................
detected frequency:  40 MHz

Component                            Vendor
UTLEON3 Processor                    AppleCore
AHB Debug UART                       Gaisler Research
AHB Debug JTAG TAP                   Gaisler Research
SVGA Controller                      Gaisler Research
AHB/APB Bridge                       Gaisler Research
LEON3 Debug Support Unit             Gaisler Research
LEON2 Memory Controller              European Space Agency
AHB ROM                              Gaisler Research
Generic APB UART                     Gaisler Research
Multi-processor Interrupt Ctrl       Gaisler Research
Modular Timer Unit                   Gaisler Research
PS/2 interface                       Gaisler Research
PS/2 interface                       Gaisler Research
Performance Counter                  AppleCore
AMBA Wrapper for OC I2C-master       Gaisler Research
General purpose I/O port             Gaisler Research
AMBA Wrapper for OC I2C-master       Gaisler Research

Use command 'info sys' to print a detailed report of attached cores

grlib>
```

Next enable the instruction and AHB trace buffers in GRMON by typing:

```
grlib> tmode both
Combined instruction/AHB tracing(0)
```

Next load the binary file with the sample program into the RAM:

```
grlib> load ../../software/utbench/demo-fib-ut/demo-fib-ut.mtram.elf
section: .text at 0x40000000, size 656 bytes
total size: 656 bytes (831.3 kbit/s)
read 47 symbols
entry point: 0x40000000

note: loaded file is an MTSPARC executable
```

You can check that the program has been loaded in the memory:

```
grlib> disas 0x40000000
  40000000   81980000    mov    0, %tbr
  40000004   01000000    nop
  40000008   01000000    nop
  4000000c   83580000    mov    %tbr, %g1
  40000010   82886ff0    andcc  %g1, 0xff0, %g1
  40000014   32800028    bne,a  0x400000b4
  40000018   01000000    nop
  4000001c   821020c0    mov    192, %g1
  40000020   81884000    mov    %g1, %psr
  40000024   81900000    mov    0, %wim
  40000028   81800000    mov    0, %y
  4000002c   3d140000    sethi  %hi(0x50000000), %fp
  40000030   bc17a270    or     %fp, 0x270, %fp
  40000034   9c27a060    sub    %fp, 96, %sp
  40000038   01000000    nop
  4000003c   01000000    nop
```

This listing can be compared with the object dump of the executed program file:

```
mtsparc-elf-objdump -D demo-fib-ut.mtram.elf
```

Run the program:

```
grlib> run
```

The program results (Fibonacci numbers) are stored in the memory. To dump the memory area that contains the results use the *mem* command:

```
grlib> mem 0x40000290 100

40000290   00000001   00000001   00000002   00000003    ...............
400002A0   00000005   00000008   0000000d   00000015    ...............
400002B0   00000022   00000037   00000059   00000090    ..."...7...Y....
400002C0   000000e9   00000179   00000262   000003db    .......y...b....
400002D0   0000063d   00000a18   00000000   00000000    ...=...........
400002E0   00000000   00000000   00000000   00000000    ...............
400002F0   00000000   00000000   00000000   00000000    ...............
```

When the XUPV5 board is connected via RS232, you should see a similar listing in your serial terminal window:

```
===== Fibonacci demo =====

index    Fibonacci number
1:       0
2:       1
3:       1
4:       2
```

```
5:      3
6:      5
7:      8
8:      13
9:      21
10:     34
11:     55
12:     89
13:     144
14:     233
15:     377
16:     610
```

The contents of the integer register file can be displayed with the *register* command:

```
grlib> register

     INS        LOCALS      OUTS      GLOBALS
  0: 00000000   00000000    0000063D  00000000
  1: 00000000   00000000    00000A18  00000001
  2: 00000000   00000000    00000000  FFFFFFFF
  3: 00000000   00000000    00000000  80000808
  4: 00000000   00000001    00000000  8000080C
  5: 00000000   00000001    00000000  00000000
  6: 50000270   0000003F    50000210  00000000
  7: 00000000   00000000    00000000  00000000

  psr: F30000E0   wim: 00000000   tbr: 40000000   y: 00000000

  pc:  400000b8  ta   0x1
  npc: 400000bc  nop
```

Commands *inst* and *hist* can be used to display the contents of the instruction trace buffer, *ahb* to display the AHB trace buffer. You should see similar results as the ones below. You can clearly identify specific parts of the microthreaded code being executed on the UTLEON3 processor.

```
grlib> inst 50
    time    address    instruction             result
    2266    4820024c   mov   %sp, %o3          [00000059]
    2267    48200250   mov   %g2, %o4          [00000090]
    2269    48200254   st    %g2, [%g1 + %g4]  [400002bc 00000090]
    2270    48200258   nop                     [00000000]
    2280    48300248   add   %o5, %sp, %g2     [000000e9]
    2281    4830024c   mov   %sp, %o3          [00000090]
    2282    48300250   mov   %g2, %o4          [000000e9]
    2284    48300254   st    %g2, [%g1 + %g4]  [400002c0 000000e9]
    2285    48300258   nop                     [00000000]
    2288    48400248   add   %o5, %sp, %g2     [00000179]
    2289    4840024c   mov   %sp, %o3          [000000e9]
    2290    48400250   mov   %g2, %o4          [00000179]
    2292    48400254   st    %g2, [%g1 + %g4]  [400002c4 00000179]
    2293    48400258   nop                     [00000000]
    2294    48100248   add   %o5, %sp, %g2     [00000262]
    2295    4810024c   mov   %sp, %o3          [00000179]
    2296    48100250   mov   %g2, %o4          [00000262]
    2298    48100254   st    %g2, [%g1 + %g4]  [400002c8 00000262]
    2299    48100258   nop                     [00000000]
    2307    48200248   add   %o5, %sp, %g2     [000003db]
    2308    4820024c   mov   %sp, %o3          [00000262]
    2309    48200250   mov   %g2, %o4          [000003db]
    2311    48200254   st    %g2, [%g1 + %g4]  [400002cc 000003db]
```

```
2312   48200258   nop                                    [00000000]
2315   48300248   add   %o5, %sp, %g2                     [0000063d]
2316   4830024c   mov   %sp, %o3                          [000003db]
2317   48300250   mov   %g2, %o4                          [0000063d]
2319   48300254   st    %g2, [%g1 + %g4]                  [400002d0 0000063d]
2320   48300258   nop                                    [00000000]
2321   48400248   add   %o5, %sp, %g2                     [00000a18]
2322   4840024c   mov   %sp, %o3                          [0000063d]
2323   48400250   mov   %g2, %o4                          [00000a18]
2325   48400254   st    %g2, [%g1 + %g4]                  [400002d4 00000a18]
2326   48400258   nop                                    [00000000]
2432   48000218   mov   %15, %16                          [00000001]
2436   4800021c   nop                                    [00000000]
2437   48000220   nop                                    [00000000]
2439   48200180   nop                                    [00000000]
3207   48200184   flush  0x0                              [00000000]
3281   48200188   nop                                    [00000000]
3282   4820018c   sethi  %hi(0x40000000), %g1             [40000000]
3283   48200190   or    %g1, 0xa4, %g1                    [400000a4]
3357   48200194   jmp   %g1                               [40000194]
3361   48200198   nop                                    [00000000]
3364   482000a4   mov   -1, %g2                           [ffffffff]
3365   482000a8   nop                                    [00000000]
3698   482000ac   flush  0x0                              [00000000]
3772   482000b0   nop                                    [00000000]
3776   482000b4   mov   1, %g1                            [00000001]
3777   482000b8   ta    0x1                               [trapped]
```

```
grlib> ahb
      time      address    type    data      trans size burst mst lock resp  hirq
      3734    400000d8    read    01000000    3    2    1    0    1    0    0000
      3738    400000dc    read    01000000    3    2    1    0    1    0    0000
      3742    400000e0    read    01000000    3    2    1    0    1    0    0000
      3746    400000e4    read    01000000    3    2    1    0    1    0    0000
      3750    400000e8    read    01000000    3    2    1    0    1    0    0000
      3754    400000ec    read    01000000    3    2    1    0    1    0    0000
      3758    400000f0    read    01000000    3    2    1    .0   1    0    0000
      3762    400000f4    read    01000000    3    2    1    0    1    0    0000
      3766    400000f8    read    01000000    3    2    1    0    1    0    0000
      3770    400000fc    read    01000000    3    2    1    0    1    0    0000
```

The UTLEON3 processor uses performance counters to measure the activity of the processor components or frequency of events that occur in the processor. Table F.2 describes the meaning of the counters, their values are in clock cycles. To show them in GRMON use the command *appleperf*:

```
grlib> appleperf count

    Counter 0: 437
    Counter 1: 76
    Counter 2: 0
    Counter 3: 0
    Counter 4: 319
    Counter 5: 76
    Counter 6: 0
    Counter 7: 0
    Counter 8: 119
    Counter 9: 51
    Counter 10: 0
    Counter 11: 61
    Counter 12: 49
    Counter 13: 0
    Counter 14: 0
    Counter 15: 0
```

```
Counter 16: 0
Counter 17: 0
Counter 18: 0
Counter 19: 239
Counter 20: 0
Counter 21: 18
Counter 22: 3
Counter 23: 101
Counter 24: 18
Counter 25: 1
Counter 26: 0
Counter 27: 0
Counter 28: 0
Counter 29: 0
Counter 30: 0
Counter 31: 0
Counter 32: 0
Counter 33: 0
Counter 34: 0
Counter 35: 0
Counter 36: 0
Counter 37: 0
```

Table F.2 Performance counters

Performance counter	Description
0	program runtime
1	*holdn* generated by I-CACHE
2	*holdn* generated by D-CACHE
3	*holdn* generated by the FP unit
4	all force NOPs
5	UTLEON3 *holdn*
6	scheduler.pick to iu3 busy
7	scheduler.push to iu3 busy
8	the number of committed instructions (pipeline valid)
9	I-CACHE grant
10	D-CACHE grant
11	regfile – write on sync port
12	regfile – write on async port
13	regfile – write on both ports concurrently
14	RAU – release register block
15	RAU – allocate register block
16	D-CACHE miss
17	*holdn* generated by the scheduler
18	*holdn* generated by the regfile
19	iu3 XC state – force NOP
20	iu3 XC state – annuled due to instruction loading latency
21	iu3 XC state – data dependencies
22	iu3 XC state – annuled due to an operation that accessed a *PENDING* register
23	iu3 XC state – working
24	iu3 DE switch
25	iu3 EX switch
26	HWFAM DMA – AMBA occupancy
27	HWFAM DMA – words downloaded from the main memory
28	HWFAM DMA – words uploaded to the main memory
29	HWFAM – n.o. queries
30	HWFAM – families created
31	HWFAM – reading from the regfile
32	HWFAM – writing to the regfile
33	HWFAM – waiting for the regfile
34	HWFAM – FIR0 – busy
35	HWFAM – FIR0 – running
36	HWFAM – FIR1 – busy
37	HWFAM – FIR1 – running

References

1. Arvind K, Nikhil RS (1990) Executing a program on the MIT tagged-token dataflow architecture. IEEE Trans Comput 39(3):300–318
2. Gaisler J Fault-tolerant microprocessors for space applications. http://www.gaisler.com/doc/vhdl2proc.pdf. Accessed September 7, 2012
3. Gaisler J (2001) The LEON processor user's manual. Gaisler research
4. Gaisler J, Catovic E, Habinc S (2007) GRLIB IP library user's manual. Gaisler research
5. Gaisler J, Catovic E, Isomaki M, Glembo K, Habinc S (2007) GRLIB IP core user's manual. Gaisler research
6. Guz Z, Bolotin E, Keidar I, Kolodny A, Mendelson A, Weiser UC (2009) Many-core vs. many-thread machines: stay away from the valley. IEEE Comput Archit Lett 8(1):25–28
7. IEEE (1994) IEEE standard for a 32-bit microprocessor architecture (IEEE-Std 1754–1994). IEEE Computer Society press, New York
8. Independent JPEG Group Independent JPEG Group. http://www.ijg.org/. Accessed September 7, 2012
9. Jesshope C (2004) Scalable instruction-level parallelism. In: Computer systems: architectures, modeling, and simulation. Springer, Berlin/Heidelberg, pp 383–392
10. Jesshope CR (2006) μTC – an intermediate language for programming chip multiprocessors. In: Asia-pacific computer systems architecture conference. Springer, Berlin/Heidelberg, pp 147–160
11. Jesshope CR, Luo B (2000) Micro-threading: a new approach to future RISC. In: Proceedings of the 5th Australasian computer architecture conference. IEEE Computer Society Press, Los Alamitos, pp 34–41
12. Kissell KD (2008) MIPS MT: a multithreaded RISC architecture for embedded real-time processing. In: Stenström P, Dubois M, Katevenis M, Gupta R, Ungerer T (eds) High performance embedded architectures and compilers. Lecture notes in computer science, vol 4917. Springer, Berlin/Heidelberg, pp 9–21
13. Kongentira P, Aingaran K, Olukotum K (2005) Niagara: a 32-way multithreaded SPARC processor. IEEE Micro 25(2):21–29
14. Saavedra-Barrera RH, Culler DE, von Eicken T (1990) Analysis of multithreaded architectures for parallel computing. In: Proceedings of the second annual ACM symposium on parallel algorithms and architectures, SPAA '90, Island of Crete, Greece. ACM, New York, pp 169–178
15. SPARC (1992) SPARC architecture manual, Version 8. SPARC International, Inc.
16. Takayanagi T, Shin JL, Petrick B, Su J, Leon AS (2004) A dual-core 64b ULTRASPARC microprocessor for dense server applications. In: Malik S, Fix L, Kahng AB (eds) Proceedings of the 41st annual design automation conference, DAC '04, San Diego, CA. ACM, New York, pp 673–677

17. The Apple-CORE Consortium Architecture paradigms and programming languages for effi-
    cient programming of multiple COREs. http://www.apple-core.info. Accessed September 7,
    2012
18. Ungerer T, Robič B, Šilc J (2003) A survey of processors with explicit multithreading. ACM
    Comput Surv 35(1):29–63
19. Waldspurger CA, Weihl WE (1993) Register relocation: flexible contexts for multithreading.
    In: Proceedings of the 20th annual international symposium on computer architecture, ISCA
    '93, San Diego, CA. ACM, New York, pp 120–130
20. Xilinx Xilinx university program xupv5-lx110t development system. http://www.xilinx.com/
    univ/xupv5-lx110t.htm. Accessed September 7, 2012

Index

A
Active queue, 64, 91
Address space identifier (ASI), 164, **183**
Allocate, 4, 5, 17, **20–22**, 34, 35, 46, 54, 55,
 61–64, 68–70, 73, 76, 77, 81, 82, 86,
 96, 120, 127, 137, 152, 168, 202
ASI. *See* Address space identifier (ASI)
Associative coupling operator, 130
Asynchronous port, **52–54**, 85, 148

B
Basic computing element (BCE), **131**, **132**,
 134, 137, 139, 144–146, 148, 150, 153,
 156
Block size, 28, 63, 68, 70, 82, 87, **120–126**,
 137, 154, 168, 185
Break, 17, **40–41**

C
Cache hit, 48, 54, **56**, **57**, 107
CCRT. *See* Continuation create (CCRT)
Child thread, 15, 36, **74**, **76**
Cleanup queue, 63, 93
Commutative operator, 130, 131, 133
Context switch, 3, 4, 45, **48–50**, 64, 65, 120,
 122, 124, 189
Continuation create (CCRT), 67, **72–77**
Create, 5, 17, **34–35**, 46, 59, 62, 72, 88, 93, 96,
 124, 127, 128, 135, 136, 140, 148, 150,
 154, 157
Create queue, 61, **63**

D
Data cache, 10, 53, 54, **56–58**, 81, 84, **97–105**,
 174, 177, 183

Data dependencies, 4, 5, **49**, 71, **112**, **113**,
 115–119, 128, 130, 157, 202
Data-flow threads, 131
D-cache miss, **48**, 115, 124, 160, 202
DCT. *See* Discrete cosine transform (DCT)
DE_SWITCH, 13, 96
Discrete cosine transform (DCT), 116, 117,
 124, 153, 198
DMA engine, 135, 137, 140, **143**, 144, 150,
 156, 198

E
Exclusive resource, 120, 122, 126
Execution profile, 75, 76, **111–119**, 157–158
EX_SWITCH, 49, 97

F
Family creation, 15, **61–62**, 69–71, 136, 166
Family identifier (FID), 15, **36–37**, 35, 37, 38,
 61, 63, 70, 71, 86, 89, 136, 137, 148,
 165
Family of threads (FT), 4, 5, **15**, 22, 24, 28, 34,
 36, 40–43, 46, 70, 71, 114, 121, 124,
 126, 130, 133, 134, 136, 144, 147, 150,
 151, 166, 170
Family table, 4, 5, 15, 19–31, 33, 35, 36, 41,
 43, 46, 61, **64**, **85–89**, 136, 151
FE_SWITCH, 49, 96
FIB, 71–72
FID. *See* Family identifier (FID)
FIR filter, **115–119**, 132, 133, 136, 152, 155,
 157, 158
Force-nop, 64, 94, 161
Fork(), 72, 74
FT. *See* Family of threads (FT)
Functional model, 48–49

M. Daněk et al., *UTLEON3: Exploring Fine-Grain Multi-Threading in FPGAs*,
DOI 10.1007/978-1-4614-2410-9, © Springer Science+Business Media, LLC 2013